Children's Environmental Identity Development

"For everyone who seeks to connect young people with the natural world, Carie Green focuses our attention on a central question: How does an environmental identity of connection and care for nature develop during infancy, childhood, and youth? She has pondered this question at length, drawing on her own research and the work of others, and presents her reflections through a combination of research reviews and observations of children in nature covering different ages, different cultures, and different settings for social and environmental learning. A broad spectrum of readers will find much that is useful here—other researchers, parents, teachers, and staff in parks, nature centers, and institutions of all kinds that bring children and nature together."

—*Louise Chawla, Professor Emerita, University of Colorado Boulder*

"Carie Green offers unique and thought-provoking insights into children's environmental identity development. Her work on a timely topic is well-researched and masterfully presented. Green not only offers information about environmental identity, but also offers guidance on what to do with that information. Her concerns and hopes for children and the natural world are evident throughout the text. I highly recommend this book to readers from multiple disciplines and with a variety of interests."

—*Ruth Wilson, Professor Emerita, Bowling Green State University, Bowling Green, OH, and Research Library Curator, Children and Nature Network*

Children's
Environmental
Identity Development

Constance Russell and Justin Dillon
General Editors

Vol. 10

The [Re]Thinking Environmental Education series
is part of the Peter Lang Education list.
Every volume is peer reviewed and meets
the highest quality standards for content and production.

PETER LANG
New York • Bern • Berlin
Brussels • Vienna • Oxford • Warsaw

Carie Green

Children's Environmental Identity Development

Negotiating Inner and Outer Tensions in Natural World Socialization

PETER LANG
New York • Bern • Berlin
Brussels • Vienna • Oxford • Warsaw

Library of Congress Cataloging-in-Publication Data

Names: Green, Carie, author.
Title: Children's environmental development: negotiating inner and outer
tensions in natural world socialization / Carie Green.
Description: New York: Peter Lang, 2018.
Series: [Re]thinking environmental education; v. 10 | ISSN 1949-0747
Includes bibliographical references.
Identifiers: LCCN 2018021040 | ISBN 978-1-4331-3200-1 (hardback: alk. paper)
ISBN 978-1-4331-3199-8 (paperback: alk. paper)
ISBN 978-1-4331-5805-6 (ebook pdf)
ISBN 978-1-4331-5806-3 (epub) | ISBN 978-1-4331-5807-0 (mobi)
Subjects: LCSH: Environmental education—Psychological aspects. |
Children and the environment. | Environmental psychology. |
Childhood development—Environmental aspects.
Classification: LCC GE70.G74 2018 | DDC 372.35/7—dc23
LC record available at https://lccn.loc.gov/2018021040
DOI 10.3726/b14279

Bibliographic information published by **Die Deutsche Nationalbibliothek**.
Die Deutsche Nationalbibliothek lists this publication in the "Deutsche
Nationalbibliografie"; detailed bibliographic data are available
on the Internet at http://dnb.d-nb.de/.

The paper in this book meets the guidelines for permanence and durability
of the Committee on Production Guidelines for Book Longevity
of the Council of Library Resources.

This book is dedicated to my four beautiful daughters:
Heidi Rainbow, Juniper Hope, Jade Rebecca, and Hallie Rose.
May the richness of all that you are in nature grow and
flourish throughout your life!

CONTENTS

ILLUSTRATIONS

ACKNOWLEDGMENTS

First and foremost, I would like to thank my lovely four daughters, Heidi, Juniper, Jade, and Hallie for your unceasing joy and inspiration! Life is full because of you! Over mountains, through valleys, along canyons, amidst mighty trees, across rivers, beside oceans, I feel so blessed to have shared the many moments together in awe of all creation. I look forward to many more moments for the rest of our lives and am so thankful to play an active role in your Environmental Identity Development (EID).

Special thanks to Connie Russell and Justin Dillon, editors of the [Re] Thinking Environmental Education series for believing in the important role EE plays in nurturing children's growing sense of self in relation to the natural world. Thank you to Anneliese Worster and Darius Kalvaitis for contributing to the initial idea of EID. Thanks to other EE research colleagues for contributing your thoughts and insight to this work.

This book could not have been possible without involvement of the wonderful children at the Bunnell House Early Childhood Lab School at the University of Alaska Fairbanks (UAF). You continue to inspire me with a richer, more creative view of the natural world. Thank you to my undergraduate student research assistants in the *Welcome too Our Forist* research project: Renee Avugiak, Gabriel Cartagena, and Katrina Bishop. Also, thanks to Paige

VonderHaar, director of the lab school, for supporting and facilitating the children's adventures in the forest, keeping parents and community members informed of forest happenings, and organizing the many logistics of our evolving research project (from assuring art supplies to arranging tree stump tables). Thank you teacher and friends, Connie Slater, Pammy Fowler, and Chasity Evanow, for believing in the value of this project and the benefits associated with allowing children to freely play and discover their forest. Thank you for your insightful comments throughout this process and loaning your classroom space and time for in-depth discussions with the children. Thank you to all the support staff (Sooyun Chi, Shannon Harvey, Mandee Cogley, and Emily Noble) as well as the practicum teacher interns (Ashley Thronsen, Desiree Dan, Haley McGlinchy, and Laura Van Amberg), your care for the children and support was essential in making the forest a welcoming place for all! And thank you to Bunnell House parents for enriching your children's lives and allowing your children to thrive and imagine in our Alaskan forest!

Thank you for those who supported the second iteration of studying children's Environmental Identity Development in an Alaskan rural community context. Thank you Robin Child, Bethany Fernstrom, and John Henry Jr. for helping to orchestrate the project; thank you to the 54 kindergarten through third grade students for being a source of inspiration; and thank you to all of the collaborating teachers, administrators, parents, and community members at Unalakleet School for making this event possible. Our field day excursion to the nature camp – sharing a ride with children in the cabs of school pick-up trucks, cascading through fresh fallen leaves of the birch and spruce forest, trudging across the tundra to pick luscious berries, and soaking in the rays of the crisp clean ocean breeze—will forever be imprinted in my memory as a source of inspiration and a true community connection.

The research presented in the content of this book was generously supported by UAF's Undergraduate Research and Scholarly Activity (URSA) as well as the Alaska Experimental Program to Stimulate Competitive Research (EPSCoR), National Science Foundation award #OIA-1208927. Continued support for studying children's EID is being provided by the UAF Biomedical Learning and Student Training (BLaST) program, funded by the National Institute of General Medical Sciences of the National Institutes of Health under Award Numbers RL5GM118990. The content in this book is solely the responsibility of the author and does not necessarily represent the official views of the National Institutes of Health.

Special thanks to all of my colleagues at the UAF and the School of Education for supporting and believing in me through this endeavor, including: Dr. Stephen Atwater, Dr. Anupma Prakash, Dr. Allan Morotti, Dr. Raymond Barnhardt, Dr. Carol Barnhardt, Dr. Cindy Fabbri, Dr. Amy Vinlove, Dr. Susan Renes, Dr. Sean Asiqłuq Topkok, Dr. Maureen Hogan, Anne Armstrong, Don Peterson, Nicole Sletterink and many others. Thanks to graduate students in my place-based education classes and introductory research courses and thanks to undergraduate students in my child development course for sharing your thoughts, ideas, and experiences of EID. I am so pleased to be *Naturally Inspired* by UAF and the Fairbanks community. And thanks to the countless colleagues and friend from around the globe for the inspiration that has led me to this in-depth exploration of children's EID.

PREFACE

This book focuses on a timely topic, Environmental Identity Development (EID) in young children and the way such identity progresses throughout the life span, drawing together theories and research from distinct fields including environmental education, education for sustainability, environmental psychology, sociology, and child development to enrich practices in early childhood environmental education and education for sustainability.[1] In an early childhood context, EID, or the natural world socialization of young children, considers not only how the natural environment affects the growth and development of young children, but also how children, as they grow and develop, shape and influence natural settings. Such childhood relations/bonds with their environment are explicitly linked to familial, sociocultural, and geographical contexts. In other words, adults (parents, caregivers and educators), siblings and peers, as well as culture play an essential role in supporting children as they develop foundational aspects of their environmental identity.

In recent years, environmental identity has gained an increasing amount of interest among scholars in environmental education and other related disciplines (Blatt, 2014; Brügger, Kaiser, & Roczen, 2011; Clayton, Fraser, & Burgess, 2011; Clayton & Opotow, 2003; Evans, Ching, & Ballard, 2012; Hinds & Sparks, 2009; Payne, 2001; Stapleton, 2015; Stets & Biga, 2003).

Environmental identity, an aspect of self-identity, considers an individual's self-concept in relation to the natural world and is generally concerned with the degree in which one is willing to act for the environment (Clayton & Opotow, 2003; Payne, 2001). While action is the resulting behavior of one who has a strong environmental identity, and has long been recognized as an outcome of environmental education (UNESCO, 1978), little has been written about how children's environmental identities emerge. In other words, such action-oriented dispositions do not appear over night, rather they are sustained and constructed through meaningful social and independent encounters with the natural world. Additionally, although there is a general consensus that formative childhood experiences in nature are significant to the identity construction of adults and adolescents involved in environmental actions (Chawla, 1999; Palmer & Suggate, 1996; Wells & Lekies, 2006), none have yet articulated the progression of environmental identity formations across the early childhood age-spectrum. In other words, theory on children's environmental identity development (EID) is generally lacking, particularly in focusing on young children (birth to eight-years-old). Thus, this book significantly contributes to the literature by specifying a framework for interpreting young children's EID, that is, children's growing sense of self in, with and for the natural world.

The EID progressions proposed in this book are derived from both psychological and contemporary sociological understandings of childhood, recognizing children as active agents, and provide a fresh lens for considering how experiences in the natural world inform a child's growing self-concept. It can be used as a theoretical lens for interpreting early childhood environmental education and early childhood education for sustainability research. It can also be applied as a tool for caregivers and educators to consider how environmental or nature-based learning experiences may be constructed to positively influence the development of children's environmental identities.

The EID model emerged from ongoing dialog between myself and other early childhood environmental education researchers focusing on child development and its relation to environmental education and education for sustainability. Specifically, while scholarship in early childhood environmental education and early childhood education for sustainability has gained an increasing amount of interest in recent years, disparities among the philosophical and theoretical views of childhood and the methodologies and methods employed in the pursuit of such research has been noted (Green, 2015b). Specifically, different orientations have emerged based on various political, sociocultural, and economical contexts around the globe. Early childhood environmental education still appears to be the term of choice in many North American and Asian nations, while early childhood education for sustainability is more prevalent in Australian and Norwegian contexts (Davis & Elliott, 2014). The former seems to be geared towards children's early learning experiences in nature, while the latter appears to be more centered on critical discourse and actions for sustainability. This book attempts to bridge these parallel, yet sometimes opposing, discourses by drawing together both traditional theories of psychosocial development and contemporary sociological understandings of childhood, as well as current research in environmental education and beyond.

Since the time that the EID model was first presented and published (see Green, Kalvaitis, & Worster, 2016), new insights have and still continue to emerge. Specifically, while the discussion of EID initially focused on an early childhood context, it became evident early on that the theory proposed in this book could be applicable to not only very young children, but also to elementary-aged children, teenagers, *and* adults. While the vignettes presented in this book are primarily positioned within an early childhood context, that is simply because this is where the bulk of my own research resides. However, some of my recent graduate students have adopted and applied EID to other contexts, including: a place-based outdoor-focused elementary school, among freshman high school students learning about their local ecology, and adolescent girls participating in a 'Girls on Ice' glacier exploration program. Additionally, I am in the process of facilitating a new study to explore how children's EID might emerge in a rural Alaska Native community context. Indeed, it is my hope to holistically extend understanding of EID in a variety of contexts and situations, recognizing that identity formation is informed by various sociocultural, political, geographical, and yes, even spiritual contexts. Perhaps the emotional and the spiritual contexts are the most important – although scarcely looked at.

The EID model also surfaced from my research on young children's special places, which revealed how important *Spatial Autonomy* and *Environmental Competency* are to children's environmental identity formation (Green, 2011; 2013; 2015a). Indeed, young children have a need to claim their own places, which serve various purposes in their lives, from places of play and exploration to places of rest and retreat. Such places provide children with spaces to explore who they are, their own preferences and beliefs.

Additionally, the EID theory presented in this book draws from findings from two recent research projects involving young children between the ages of three and nine years of age. The first titled: *Welcome too Our Forist* project (named by the children) occurred during the summer of 2015 and involved thirty-one children between the ages of three to six years. The second project titled: *Children's Environmental Identity Development in an Alaska Native Context* involved fifty-three children between the ages of five to nine years and occurred in early fall 2016. The Institutional Review Board at the University of Alaska Fairbanks approved both research projects. Additionally, parental consent and child assent was obtained prior to participation. Names of children and adults are pseudonyms.

The two projects are described below.

The *Welcome too Our Forist* research project

The primary goal of the *Welcome too Our Forist* research project was to explore research methods to engage young children as active researchers of their own experiences in a forest. The children, with their teachers and researchers, visited the same patch of forest located in a non-rural setting eleven times for approximately an hour over an eight-week summer period. The teachers supported the children in exercising agency in choosing where they wanted to explore, what activities they wanted to engage in, and by permitting risk-taking activities (climbing trees, hanging on branches). However, they were always nearby to monitor and ensure children's safety while in the forest. While the intent of the project was to explore methods for engaging young children as active researchers in all aspects of research, including posing researchable questions, choosing data collection methods, collecting and analyzing data, and presenting and disseminating their own findings, the study also provided important insight into children's own perspective of their environmental identity formation. The children collected their own data of the topics that they choose, including their experiences of bugs, sticks, trees, X-marks-the-spot, and castles, forts, and houses in the forest. Utilizing the Sensory Tour method, the children wore cameras to collect video of their forest explorations, took photographs with iPads, created art, built models, and engaged in role-playing to explore these topics. The children then analyzed and interpreted their findings through video-stimulated group discussions and book making explorations where each child had the opportunity to richly describe and recall his or her own experiences using photos, words, and art. (Please refer to Green (2016) and Green (2017a) for a detailed description of the methods used in the project. Research methods for studying children's EID will be examined in Chapter 6 of this book.)

The *Welcome too Our Forist* research project culminated with the children's final presentation in the forest giving them an opportunity to show and tell about their engagement in the forest environment. Findings from the project, primarily in the form of Sensory Tour transcriptions, will be woven throughout this book and referred to as the *Welcome too Our Forist* research project, following the authentic spelling given by children. This serves as a reminder that EID emerges from the inner life of a child and recognizes children as active agents of culture and change.

Children's Environmental Identity Development in an Alaskan Rural Context

The second project, *Children's Environmental Identity Development in an Alaska Rural Context*, specifically focused on children's EID in a rural village context (Green, 2017b).[2] While the *Welcome too Our Forist* research project occurred in a forest near an Alaskan "urban" setting, this study was conducted in a small rural Alaska Native village of approximately 696 people located in Northwestern Alaska along the coast of the Bering Sea (U.S. Census Bureau, 2016). Approximately 75% of the village residents are of Alaska Native, primarily Iñupiat, descent. The children represented in the project were similar to the village demographic. The isolated village is "off the road system" over 500 miles from the Alaskan urban hubs, and only accessible by plane, boat, dog sled, or snowmobile. Due to high cost of transportation and limited access, modern conveniences and supplies (including food) are limited. Many rural Alaskan residents still engage in cultural subsistence practices.

The study followed a participatory phenomenology. The children participated in a number of research activities over the course of five days to learn about their experiences and perspectives of their environment. Data was collected with children through draw and write prompts, children's photography, and video footage of Sensory Tours. The Sensory Tours, discussed in more detail in the next section, occurred at a forested nature camp located approximately 20 minutes away from the village. Children also participated in in Sensory Tours while picking berries on the tundra. These tours were self-guided and lasted between 5–20 minutes. Adults accompanied two or more children on their tours.

Sensory Tours

During both research projects children were invited to participate in Sensory Tours, where they wore a small video camera (GoPro) around their forehead while "touring" their environment. "The camera goes where a child goes, sees what a child sees, and hears what a child hears. As such it provides a unique lens for researchers to view the world through the perspective of a child" (Green, 2017b, p. 311). Advantages, challenges, and opportunities in using the Sensory Tour method with children are discussed in more detail in Chapter 6 of this book. However, for now, it is important to note that video transcriptions of children's Sensory Tours from both research projects will be woven throughout the chapters of this book to illustrate different aspects of and characteristics of children's EID. Each tour will be foregrounded by indicating the project in which it was collected. Transcriptions of children's sensory tours will be indented in the text. The statements made by the children as well as the names they attribute to various phenomenon will be italicized within the book in order to draw attention to children's voices and their agency in their EID. Furthermore, transcriptions of children's Sensory Tours will include descriptions of children's bodily actions, activities, and the environmental settings in which such tours took place. While some interpretation will be provided to contextualize each progression of children's EID, readers are encouraged to further interpret the meaning of children's EID by reflecting on their own personal and professional experiences.

Notes

1. The EID theory was first proposed in Green, C., Kalvaitis, D., & Worster, A. (2016). Recontextualizing psychosocial development in young children: A model of environmental identity development. *Environmental Education Research, 22*(7),1025–1048. https://www.tandfonline.com/doi/abs/10.1080/13504622.2015.1072136 Content of the article reproduced with permission from Taylor and Francis.
2. Some examples from the *Children's Environmental Identity Development in an Alaska Rural Context* research project were previously published in the article: C. Green (2017). Children's Environmental Identity Development in an Alaska Native Rural Context. *International Journal of Early Childhood, 48*(3), 303–319. https://link.springer.com/article/10.1007%2Fs13158-017-0204-6. Materials have been adapted and reprinted with permission of Springer Nature.

References

Blatt, E. (2014). Uncovering students' environmental identity: An exploration of activities in an environmental science course. *The Journal of Environmental Education*, 45(3), 194–216.

Brügger, A., Kaiser, F. G., & Roczen, N. (2011). One for all? *European Psychologist*, 16(4), 324–333.

Chawla, L. (1999). Life paths into effective environmental action. *The Journal of Environmental Education*, 31(1), 15–26.

Clayton, S., Fraser, J., & Burgess, C. (2011). The role of zoos in fostering environmental identity. *Ecopsychology*, 3(2), 87–96.

Clayton, S., & Opotow, S. (2003). *Identity and the natural environment*. Cambridge, MA: MIT Press.

Davis, J., & Elliott, S. (2014). *Research in early childhood education for sustainability: International perspectives and provocations*. New York, NY: Routledge.

Evans, E., Ching, C. C., & Ballard. H. L. (2012). Volunteer guides in nature reserves: Exploring environmental educators' perceptions of teaching, learning, place and self. *Environmental Education Research*, 18(3), 391–402.

Green, C. (2011). A place of my own: Exploring preschool children's special places in the home environment. *Children, Youth, and Environments*, 21(2), 118–144.

Green, C. (2013). A sense of autonomy in young children's special places. *International Journal of Early Childhood Environmental Education*, 1(1), 8–33.

Green, C. (2015a). "Because we like to": Young children's experiences hiding in their home environment. *Early Childhood Education Journal*, 43(4), 327–336.

Green, C. (2015b). Towards young children as active researchers: A critical review of the methodologies and methods in early childhood environmental education research. *The Journal of Environmental Education*, 46(4), 207–229.

Green, C. (2016). Sensory tours as a method for engaging children as active researchers: Exploring the use of wearable cameras in early childhood research. *International Journal of Early Childhood*, 48(3), 277–294.

Green, C. (2017a). Four methods for engaging young children as environmental education researchers. *International Journal of Early Childhood Environmental Education*, 5(1), 6–19.

Green, C. (2017b). Children's environmental identity development in an Alaska Native rural context. *International Journal of Early Childhood*, 49(3), 303–319.

Green, C., Kalvaitis, D., & Worster, A. (2016). Recontextualizing psychosocial development in young children: A model of environmental identity development. *Environmental Education Research*, 22(7), 1025–1048.

Hinds, J., & Sparks, P. (2009). Investigating environmental identity, well-being, and meaning. *Ecopsychology*, 1(4), 181–186.

Palmer, J. A., & Suggate, J. (1996). Influences and experiences affecting the pro-environmental behaviour of educators. *Environmental Education Research*, 2(1), 109–121.

Payne, P. (2001). Identity and environmental education. *Environmental Education Research*, 7(1): 67–88.

Stapleton, S. (2015). Environmental identity development through social interactions, action, and recognition. *The Journal of Environmental Education, 46*(2), 94–113.

Stets, J. E., & Biga, C. F. (2003). Bringing identity theory into environmental sociology. *Sociological Theory, 21*(4), 398–423.

United Nations Educational, Scientific and Cultural Organization (UNESCO). (1978). Intergovernmental Conference on Environmental Education: Tbilisi (USSR), October 14–26, 1977. Final Report. Paris: UNESCO.

Wells, N., & Lekies, K. (2006). Nature and the life course. *Children, Youth and Environments, 16*(1), 1–24.

A MODEL FOR ENVIRONMENTAL IDENTITY DEVELOPMENT

A Twilight Dance in the Snow[1]

Dusk in the far north comes early in the wintertime, by three in the afternoon it is dark. If one does not act quickly during the short daylight hours the window for outdoor play has withered away. Or has it?

"What would you like to do during the winter break?" I asked my three young daughters during a dinner conversation. My 7-year-old, Heidi, suggested a museum, my 5-year-old, Juniper, a hike, and my 3-year-old, Jade just giggled.

My own bias showed through their suggestions. A hike? How about a late night stroll in the woods? Enthusiasm was apparent and so after stowing away the food and dishes we proceeded with putting on the layers of winter gear appropriate for outdoor adventure in the Arctic. Missing gloves, struggle with boots, and limited time caused some apprehension in our decision to go outdoors. However, we overcame the struggles and emerged outside. The dark was calm and chill; giant snowflakes softly meandered to the ground, caressing our rosy cheeks and brows. It was a perfect night for exploration!

Placing Jade in the sled we proceeded on toward the dark cluster of trees, the leafless birch glowing white in the night and the towering spruce covered with clumps of thick glistening snow. "*Which way do we go?*" Heidi asked as we entered into the trees.

"That way." I answered pointing towards the wider trail, the perfect width for a family parade. "You lead the way," I suggested to Heidi.

"*I am scared,*" she replied.

"It's not scary—it will be fun." I reassured her.

Drawing strength from her countless experiences in nature, she proceeded ahead slowly and cautiously with her small flashlight. Gaining comfort her speed quickened until she was eventually running through the deep shimmering snow. Juniper cowered behind, a little more apprehensive than her sister; she clung to her father for comfort. Her father reassured her and she eventually ventured out independently on her own, following in the boot prints of her older sister. Jade remained quite content bundled warm and safe in the sleigh.

Along the path the oldest paused, pointing her light into the shadows of the black trees; everything appeared a bit differently in the dark. We all briefly stopped and silently turned out the flashlights to fully take in the setting. How amazing were the reflections from the moon, beaming bright on the iridescent snow. The small and great trees alike stood as silent silhouettes like guardians of strength in the night. All was crisp and quiet and fresh and peaceful. We continued onward. The two older ones became braver, gaining a new awareness and sense of self in the forest at night.

Eventually we came upon a wide clearing where the snow was nearly two feet thick. A winter wonderland, the children's' energy soared to new heights as they ran in circles, carving snow angels and furrows through the soft powder. Jade awakened too, curious about the new setting. I reached out my hand to her; she was still a bit apprehensive about stepping out of the sled so I modeled how fun it was to carve a path where none existed. She smiled excitedly, stood up, and danced her own pattern through the flakes, forming her own connection through experience in the snow. We then intertwined our paths together in a shared pattern of a twilight dance in the snow.

Their father took the sled to the top of the nearby hill, racing to the bottom. The older two took their turn climbing up and sliding down carving a slippery white trench. The littlest indicated she would like a turn; looking up she said, "*Mama hold me.*" I took her hand and helped her find her way through the thick powder to the top of the hill. At the top we rode down together; I held her close providing reassurance and comfort during our novel nightfall adventure.

A Model for Environmental Identity Development (EID)

The vignette, A *Twilight Dance in the Snow*, shows the diverse ways in which children develop their environmental identities alongside parents, caregivers, educators, siblings, and peers during the early years of life. Starting with *Trust in Nature* as a foundation, children are propelled to find their own sense of *Spatial Autonomy* and develop *Environmental Competencies* through sustained meaningful encounters with the natural world. In the vignette above, each of my children appeared to progress in different ways through various stages of the Environmental Identity Development (EID) model (see Figure 1.1 below).

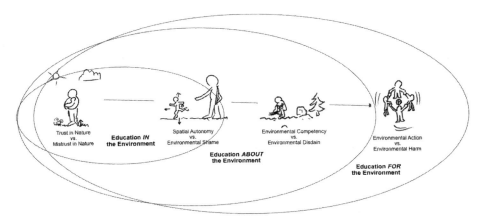

Figure 1.1. Environmental Identity Development Model © Carie Green.

For instance, as we ventured into the forest at night all of my children began at the *Trust in Nature* vs. *Mistrust in Nature* stage. This was evident when Heidi stated, "*I am scared.*" Juniper also revealed her initial discomfort or *Mistrust in Nature* by clinging apprehensively to her father. Jade, the youngest, was perhaps the most anxious, choosing to stay within the comforts of the sled. However, it did not take long for Heidi to run ahead independently and find *Spatial Autonomy*, or her own sense of place, in the dark night. She experimented by turning out the flashlight and taking in the "new" setting. Likewise, after finding reassurance from her father, Juniper also ventured along the trail, following the lead of her older sister. Then in coming into the wide-open snowfield the two displayed *Environmental Competency*, with the skills and confidence to use the environment to carry out their own goals and enrich their experience. Heidi lay down to make a snow angel and Juniper ran fervently, carving new furrows in the snow. They both took a turn sliding down a nearby hill.

Jade also progressed through the various stages of the EID model, but much differently than her sisters. She possessed a less developed sense of *Trust in Nature*, which prohibited her from independently achieving *Spatial Autonomy* or *Environmental Competency*. However, with the comfort and security provided by her family, she took small steps in carving out her own space in the snow and gaining a shared sense of *Environmental Competency* through a joint sleigh ride.

By beginning this book with the vignette, *A Twilight Dance in the Snow*, I argue that *Trust in Nature* plays an essential role in children developing a positive and healthy environmental identity. Additionally, family is recognized as a significant context within which a child's environmental identity emerges. However, not all children have opportunities to build and establish trusting relationships with nature through shared and supported family adventures. Therefore, formative encounters with the natural world must also be supported and nurtured by educators, extended family, and other adults actively involved in the lives of young children. Furthermore, in recognizing children as active agents in the construction of their own lives and the lives of those around them, peer and sibling culture also influence children's growing sense of self in the environment.

What is Environmental Identity?[2]

Environmental identity is defined as an aspect of an individual's self-concept in relation to the natural world. As explained by Clayton (2003):

> Environmental identity is one part of the way in which people form their self-concept: a sense of connection to some part of the nonhuman natural environment, based on history, emotional attachment, and/or similarity, that affects the way in which we perceive and act toward the world; a belief that the environment is important to us and an important part of who we are. (pp. 45–46)

Further, environmental identity can be understood as a "collective" identity formed within diverse cultural and social situations (Clayton, 2003, p. 46). However, environmental identity may vary among individuals, ranging on a scale from low to high, with a stronger identity resulting in stronger motivations and commitments towards "personal, social, and political behaviors" (Clayton, 2003, p. 46).

Interest in environmental identity has surged in recent years (e.g., Blatt, 2014; Evans, Ching, & Ballard, 2012; Prévot, Clayton, & Mathevet, 2018; Sharma-Brymer, Gray, Brymer, 2018; Stapleton, 2015; Tugurian & Carrier, 2017; Williams & Chawla, 2016). Stemming from Payne's (2001) assertion of a "major 'lack' in the discourse" (p. 67) on identity in environmental education research, identity becomes particularly important for considering how an individual's sense of self directs their actions and behaviors "for being for the environment" (Payne, 2001, p. 67). Similarly, Stets and Biga (2003) argued that because identity theories recognize larger social structures, they are better suited towards predicting environmental behaviors across situations than theories on how an individual's attitude towards the environment affects their behaviors. In other words, identity considers attitudes along with the emotive, moral, and social cognitions related to an individual's sense of self; these cognitions influence the way an individual acts and behaves across various contexts and situations (Prévot et al., 2018).

Others have used the term ecological identity to refer to the "ways peoples construe themselves in relationship to the earth" (Thomashow, 1995, p. 3). Ecological identity has also been used to refer to a political orientation positioned as a form of a social movement (Light, 2000). This is similar to Clayton and Opotow's (2003) interest in interrogating an individual's environmental identity positioning in relation to his or her orientation to act for environmental issues.

Place identity is a related theory that recognizes human-environmental relations. Proshansky and Fabian (1987) defined place identity as a substructure of self-identity, comprised of:

> cognitions about the physical environment that also serve to define who the person is ... The cognitions are represented as thoughts, memories, beliefs, values, ideas, preferences, and meanings relating to all the important settings of the person's daily life, past as well as present ... Place-identity cognitions monitor the person's behavior and experience in the physical world. (pp. 22–23)

Place identity as well as place attachment theories consider the significance of human bonding to places and how positive (or negative) identifications with place drive environmental behaviors (Manzo & Devine-Wright, 2014). Place identity is concerned with specific locales (i.e., this farm, this hill, this community). As Holmes (2003) explains, "Our experiences is never of 'the earth' as an actual whole, but of some particular place on the earth, a place defined by ... physical boundaries" (p. 30). Such spaces or locales may range in scale from small to large and can include both natural and human-built environmental features, such as a piece of furnishing or a city landscape (Seamon, 2014). In this way, place identity, although it may include natural places or objects, is not specifically focused on an individual's sense of self in relation to the natural world. Thus, I use EID in this book to signify a child's developing self-concept in relation to the natural environment. With that said, I also recognize that the EID of young children is likely to be situated in small-scale and familiar places (i.e., outdoor play areas, nature trails). As a child's environmental identity develops, cognitions and associations with the natural environment may shift from those that are familiar and concrete to those that are more universal and abstract.

Thus far, research on environmental identity has mainly focused on adults (Brügger, Kaiser, & Roczen, 2011; Clayton, Fraser, & Burgess, 2011; Evans et al., 2012; Hinds & Sparks, 2009; Prévot et al., 2018). Studies involving children have been limited and primarily center on middle childhood and adolescent experiences (Blatt, 2014; Gebhard, Nevers, & Billmann-Mahecha, 2003; Kals & Ittner, 2003; Müller, Kals, & Pansa, 2009; Stapleton, 2015). Studies have focused on specific aspects of environmental identity by examining how children attribute human qualities to natural objects (e.g., trees) to better understand themselves (Gebhard et al., 2003). Others have focused on emotional, cognitive, and social variables of children and youth's environmental identities and how such variables relate to an individual's commitment to protect the environment (Blatt, 2014; Kals & Ittner, 2003; Müller et al., 2009; Stapleton, 2015). Perhaps, one reason that there has been little research on young children's environmental identity is because environmental advocacy, which has been a prominent aspect of environment identity theory, is not as prevalent during the early years. Indeed, environmental behaviors in early childhood are often characterized by exploration and discovery (Hart, 1979; Wilson, 1996). Additionally, some traditional child development theories have characterized young children as predominately egocentric, or focused on self (McDevitt & Ormrod, 2013). Such orientations have attributed more sophisticated social cognitions of relating self with other, humans and environmental features, to the later part of early childhood and the middle childhood years.

Closely related to environmental identity, Wilson (1996) theorized the "ecological self" of young children and how self-cognitions, in relation to the environment, may evolve from childhood to adulthood. She explains that young children have a "unique affinity" for the natural environment and experience it in a "deep and direct manner … as a factor and stimulator" (p. 121).

However, she notes that this affinity tends to decrease as children get older. Drawing from Sebba's (1991) work on the *Landscape of Childhood*, Wilson (1996) explained that many families in Western or urbanized cultures tend to alienate children from natural environments, replacing direct sensory-rich nature experiences with encounters with human-made replicas or models. As Sebba (1991) noted, removing children from the natural environment "is accompanied not only by physical but also by psychological separation" (p. 414). Indigenous scholars concerned with the displacement of Native worldviews or identities have also noted the psychological separation between humans and their landscape (Kawalgey, 2006; Lowan-Trudeau, 2013). As a result of a nature disconnection, children are developing "critical and analytical" dispositions towards the natural environment rather than "adaptive and sympathetic attitude(s)" (Sebba, 1991, p. 414). Indeed, this isolation from nature and its impact on the health and well being of children has been thoroughly discussed in Louv's (2008) *Last Child in the Woods* and by "children in nature" advocates across the world (Children & Nature Network, 2014). Part of this disconnect, or children's "nature deficit disorder," has been attributed to too many structured extracurricular activities and not enough time outside as well as parents' fear of the risks associated with unstructured play in nature (Louv, 2008). In addition, research also suggests that avoidance of nature experiences in early childhood may impact the way in which older adolescents relate to and experience nature. In other words, children who have had limited or adverse experiences in nature may display discomfort, disgust, and anxiety in natural environments (Bixler, Carlisle, Hammitt, & Floyd, 1994; Bixler & Floyd, 1997).

In contrast, children's participation in nature has been shown to lead to pro-environmental attitudes and behaviors in adulthood (Wells & Lekies, 2006). As well, a growing body of literature in early childhood environmental education and early childhood education for sustainability points towards a progression of EID. For instance, Barratt, Barratt-Hacking, and Black (2014) propose that young children need to first get to know, understand, and connect with natural environments. They propose the necessity of "spending sustained periods of time outdoors" (p. 228) and interacting with it as an important prerequisite for protecting and acting for the environment. Similarly, Chawla and Rivkin (2014) argue that play in nature allows children the opportunity to become familiar with the natural environment and develop competencies:

> It [nature play] is filled with graduated challenges that enable children to reach beyond past accomplishment to the next level of possibilities – a higher tree branch

that they could not grasp before but find accessible today, a heavier stone to lift, a larger log to roll, a more distant path to navigate. (p. 252)

Alternatively, Ritchie (2014) proposes a "critical place-based orientation" in early childhood education for sustainability, explaining how, "dialogical interaction with both Indigenous peoples and the local place itself, is seen as a source for interpreting ways of caring deeply for our planet, positioning humans alongside local ecologies as 'co-habitors' of the earth" (p. 49). In this way, Ritchie (2014) extends the importance of considering value development and the spiritual well being of children as well as their physical and emotional well being experienced through "sensory, storied entanglement within the inter-relational agency of other animals, plants, insects, and the rest of the [more-than-human] world around us" (p. 50). Likewise, Hägglund and Johansson (2014) posit belonging, recognizing oneself as part of a "world shared with others" (p. 24) as an essential value in early childhood education for sustainability discourse.

Thus, the literature suggests that time spent in nature is essential to the development of environmental competencies and that establishing a sense of belonging and deeper relations with place and the more-than-human environment is essential to promoting pro-environmental values and behaviors. However, thus far the literature has primarily focused on nature-based curriculum and pedagogy as well as promoting critical "transnational" dialogue on the ethics and values surrounding early childhood education for sustainability (Davis & Elliott, 2014). To my knowledge, none have focused on the psychosocial progressions of children's EID, and how such development is nurtured and sustained through environmental education contexts.

Bridging Psychological and Sociological Theories of Childhood

The EID model presented in this book is supported by psychological and contemporary sociological understandings of childhood, particularly in recognizing children as active agents, and provides a fresh lens for considering how experiences in the natural world inform a child's growing self-concept. The EID model draws from Erikson's (1950; 1972; 1980) psychosocial development theory. Erikson recognized the inherent autonomy of children who, at a very young age, are constructing their own meanings of the world. In this way, agency, as recognized in contemporary sociological theories (Corsaro, 2015; Qvortrup,

Corsaro, & Honig, 2009), is a fundamental part of Erikson's psychosocial developmental theory and is thus recognized in the EID model. Therefore, children are acknowledged as competent social actors "active in the construction of their own lives, and the lives of those around them and of the societies in which they live" (James & Prout, 1990, p. 8). In other words, children are not viewed as "human becomings" (Lee, 2001, p. 7); rather they are viewed as active agents and contributors to culture and change (Corsaro, 2015). Thus, the various stages of the EID model draw from research and theories that recognize children's perspectives and honor children's participatory rights and learning about what's important to them (United Nations, 1989, 2005).

Erikson's Psychosocial Stages of Development

In starting with a basic review of Erikson's framework, it is important to note that Erikson (1972; 1980) recognized both biological and environmental implications of children's development, as well as children's own initiative to learn. He proposed that children progress through a series of psychosocial dilemmas (stages) in the development of their identity. Progression through each stage is thus determined by children's success in overcoming "outer and inner" conflicts, or tensions, attributed to healthy development (i.e. *Autonomy vs. Shame and Doubt*) (Erikson, 1980). Thus, progress through each successive stage is determined by "weathering" conflicts, "emerging and reemerging with an increased sense of inner unity, with an increase in good judgment, and an increase in the capacity to do well" (Erikson, 1980, p. 52). Erikson (1980) goes on to explain that "each item of the healthy personality" is "systematically related to all others" (p. 54). For instance, if a child fails to establish a trusting relationship with his or her immediate caregiver during infancy, he or she may have difficulty forming intimate relationships later in life. With that said, it should also be noted that an overlap exists between the various stages, with each conflict emerging "in some form before 'its' decisive and critical time" (Erikson, 1980, p. 54). Thus, Erikson's model provides a degree of fluidity; children, with adult guidance, may need to revisit previous stages in order to refine competencies that may be lacking in his or her healthy identity development. In other words, although the model itself is presented through a series of age-related developmental stages, Erikson recognized that children's identities unfold in diverse ways through varied contexts and experiences. However, positive identity development occurs only when a child is able to successfully overcome the foundational dilemmas presented at each stage. If a dilemma is not successfully mastered, then a child may seek alternative measures to gain competencies that may be lacking in his or her healthy sense of self.

In the foundational stage of *Trust* vs. *Mistrust* (birth–18 months), Erikson proposed that it is during this stage that infants learn to trust adult caregiver(s) to provide for their needs. This includes basic physical needs such as satisfying hunger and changing an uncomfortable diaper as well as socio-emotional needs such as consistent comfort and affection. When care from adult(s) is inconsistent, inattentive, or abusive, infants develop mistrust and begin to perceive the world as an unsafe and unpredictable place.

In the second stage of *Autonomy* vs. *Shame and Doubt* (1–3 years), toddlers become more coordinated in their motor skills, they gain competencies in their abilities to move their bodies and manipulate objects in their environments. With this independent mobility, toddlers start to recognize themselves as autonomous individuals and become more capable of satisfying their own needs (e.g., dressing themselves and learning how to use the toilet). When caregivers encourage toddlers to practice these self-sufficient skills, toddlers develop an ability to tackle tasks and problems on their own. On the other hand, when support is lacking or caregivers become impatient, toddlers may experience feelings of shame and doubt and become discouraged in approaching tasks independently.

In the third stage of *Initiative* vs. *Guilt* (3–5 years), children's growing sense of autonomy evolves into initiative, the undertaking and planning of tasks. Here a child may exercise his or her own ideas through play and other activities. For instance, young children may role-play real or imagined scenarios in discovering who they are as individuals. When caregivers foster and support children's creative ideas, children gain a sense of excitement and initiative to try new things. On the other hand, if they are made to feel incompetent about their ideas or initiatives, children may develop a sense of shame and guilt about their abilities.

In the fourth stage of *Industry* vs. *Inferiority* (6 to 10 years), children transition into primary school, and their initiative shapes into a sense of industry driven by academic achievement, pleasing peers, teachers, and adult authority figures. With support and encouragement, children attempt and achieve tasks set before them and pursue new interests and challenges. In contrast, when children are overly critiqued and made to feel that their abilities do not measure up, they may develop a sense of inferiority coupled with feelings of doubt in their abilities to conquer challenging tasks on their own.

At the heart of Erikson's theory is the importance of young children forming trusting relationships and bonds with adult caregivers and/or educators who support their identity development. Healthy interactions with adult figures (i.e., parents, caregivers, and teachers) are essential to children's progression through the psychosocial stages. Therefore, in proposing strategies to extend Erikson's theory within an environmental education context, the role that adults play in helping children develop healthy relations with the natural world is significant.

In considering Erikson's stages, it should also be noted that his theory has been criticized for being Western-biased. For instance, some suggest that autonomy is an individualistic concept and that some collectivist and community-oriented cultures may discourage independence and encourage more reliance on family members and community ties (Rasmussen, 2009). While the level in which independence is encouraged and supported may vary among families and cultural value systems, Erikson's theory is nonetheless useful in conceptualizing how children develop their *own* relationships with their environment independently and through social interactions, while also emphasizing the essential role in which adults and other children play in that process, as noted in each of the stages.

Psychosocial Development in an Environmental Education Context

While Erikson's theory has been examined and built upon by human development scholars (Dunkel & Sefcek, 2009; Lerner, 2002), it has not yet been applied to an environmental education context. Just as other interpersonal developmental theories such as attachment theory (Ainsworth, Blehar, Waters, & Wall, 1978) have been useful for proposing theories on human-place attachments (see Scannell & Gifford, 2014), Erikson's psychosocial development theory provides a good starting point for considering how experiences in, about, and for the natural world are integral for young children's healthy EID.

A progression for EID is presented below, recognizing the early years as a significant time when aspects of one's environmental identity are formed. While the EID model proposes "ideal" timeframes for when certain environmental cognitions can emerge, it also acknowledges that missing cognitions need, and should, be revisited and enhanced through sustained, meaningful experiences with nature throughout a person's lifetime.

Foundation: Trust in Nature vs. Mistrust in Nature

Beginning with *Trust in Nature* vs. *Mistrust in Nature*, feelings of assurance and comfort in nature vs. feelings of anxiety and discomfort are fundamental to a child's ability to progress along the continuum of healthy EID. Through consistent and sustained encounters with the natural world, children have the opportunity to see, smell, hear, touch, and taste their environments. Such sensory-rich experiences are paramount to identity formation as the feel of the wind, the smell of the trees, and the taste of the air become inscribed into a child's very being. Social interactions with adults and peers play an important role; caregivers and educators can talk to them about the natural surroundings, share in their sense of wonder, and encourage children to form trusting bonds with places, beings, and objects during each encounter. The goal is to facilitate a child's sense of *Trust* vs. *Mistrust in Nature*, meaning that it is essential that adults meet children's needs by ensuring physical and psychological comfort during nature experiences (e.g., making sure a child is clothed properly and supported when encountering new things). In this way, it is also essential that caregivers/educators facilitate a healthy respect for nature's inherent dangers. This, in turn, helps a young child gain comfort in nature, while also recognizing that there are certain aspects in nature of which to be cautious or to avoid (e.g.. a prickly cactus, a bear).

First Progression: Spatial Autonomy vs. Environmental Shame

A trusting bond with nature encourages children to venture out, independently or collectively with others, to achieve *Spatial Autonomy*. In this way, children discover their own sense of place in the natural world; this, in turn enables them to explore attributes of their environmental identity. Nature in this sense, becomes a mirror in which to view one's self, a special rock becomes an extension of self, providing comfort and reassurance to process thoughts, feelings, values, and beliefs. Contrary to gaining *Spatial Autonomy* are feelings of doubt or *Environmental Shame*, where a lack of comfort and security may cause children to withdraw from nature experiences. Such negative cognitions may occur when adults discourage or prohibit children's independent exploration in the outdoor environment. Young children may initially stay within close proximity to adults, but as children's level of comfort matures they venture out further on their own. Caregivers and educators need to allow for a certain level of risk-taking behaviors in children's engagement with nature. It is through such risk and independent experiences that children gain a sense of *Spatial Autonomy* and confidence in securing *Environmental Competency* described in the next progression.

Second Progression: Environmental Competency vs. Environmental Disdain

Healthy cognitions connected with discovering a sense of *Spatial Autonomy* provide children with opportunities to acquire *Environmental Competency* gained through experimentation and creative innovations, which enrich children's experiences in nature. Fantasy play and experimentation provide children with opportunities to exercise their creativity, and emerging critical thinking and problem-solving skills to symbolically represent objects or scenarios (e.g., using a stick as a spoon, creating a "house" in a bush, building a bird's nest). In contrast, a lack of opportunities to take initiative in nature could lead to *Environmental Disdain*, or cognitions of contempt that separate children from the natural world. Instead of using the environment to fulfill one's sense of purpose, children may become indifferent to their natural surroundings, instead developing competencies associated with aspects of the material world. Caregivers and educators can support children's development of *Environmental Competency* by furthering understanding and applicability of ecological concepts related to children's interests.

Third Progression: Environmental Action vs. Environmental Harm

A strong sense of *Trust in Nature*, *Spatial Autonomy* with nature, and *Environmental Competency* to work with nature to fulfill a sense of purpose motivates children to exercise *Environmental Action*. In this way, children's previously acquired self-cognitions (i.e. values, knowledge, and care for the environment) can be applied towards environmental stewardship or creating a more sustainable future. The scale of such actions will likely be situated within familiar social and environmental contexts. In contrast, children who lack in their development in any one of the preceding environmental identity attributes may develop an environmental identity that is ignorant of or disregards the natural world. This, in turn, may lead to the establishment of dispositions or behaviors that promulgate *Environmental Harm*.

Environmental Action is the goal of a healthy EID. Caregivers/educators play an essential role in modeling and helping children consider ecological values and behaviors, while also promoting inquiry and action in response to environmental challenges. Peer culture also influences children's EID. In this way, children are recognized as competent social actors, active in the construction of their own lives and the lives of those around them. Strong environmental identity cognitions can lead children to initiate action with peers, independent from adults. However, weak environmental identity cognitions may prematurely expose children to exercise *Environmental Actions* that they may not be developmentally ready to take on.

Education *in/from*, about, and *for* the Environment

In considering an integrated model of EID, it is useful to draw upon Lucas' (1979) model of education in/from, about, and for the environment (further applied by Palmer & Suggate, 1994). Education *in* and *from* the environment is mostly concerned with directing experiences outdoors or in nature, thus inspiring a sense of wonder and relationship with the nature world. Education *about* the environment primarily focuses on building environmental knowledge, understandings, and awareness of ecological processes. Education *for* the environment is mainly directed towards promoting conservation behaviors and actions for sustainability.

Healthy EID presumes a holistic approach, recognizing all three types of learning as significant to children's environmental identity formation; however, certain types of environmental education experiences may become the focus point for children's learning during different periods. Table 1.1 provides a basic sequence for emphasizing experiences *in/from*, *about*, and *for* the environment during the different progressive stages of EID.

Negotiating Tensions

In considering EID, children are faced with negotiating outer environmental dilemmas and personal inner tensions, between *Trust in Nature* vs. *Mistrust in Nature*, *Spatial Autonomy* vs. *Environmental Shame*, *Environmental Competency* vs. *Environmental Disdain*, and *Environmental Action* vs. *Environmental Harm*. Often such dilemmas are overcome in social contexts; in other words, adults and peers play a significant role in supporting children as they navigate these tensions and progress in their EID. In the foundational stage of *Trust in Nature* vs. *Mistrust in Nature*, the goal is to support children in developing a sense

Table 1.1. EID and Education In/From, About, and For the Environment.

Developmental Stages	IN/FROM	ABOUT	FOR
Trust in Nature	Experiences IN the Environment with primary caregiver(s)	Early promotion of Education ABOUT the Environment with primary caregiver(s)	Early promotion of Education FOR the Environment with primary caregiver(s)
Spatial Autonomy	Education IN the Environment with growing independent exploration	Early promotion of Education ABOUT the Environment mainly with caregiver(s) and family members	Early promotion of Education FOR the Environment, mainly with caregiver(s) and family members
Environmental Competency	Continued focus on Education IN the Environment through independent and social contexts	Experiences ABOUT the Environment through independent and social manipulation	Early promotion of Education FOR the Environment, with peers and adult caregivers
Environmental Action	Continued focus on Education IN the Environment Experiences through independent and social contexts	Continued focus on providing Experiences ABOUT the Environment through inquiry and exploration	Actions FOR the Environment promoted in familiar social contexts and settings

Table originally published in Green, C., Kalvaitis, D., & Worster, A. (2016). Recontextualizing psychosocial development in young children: A model of environmental identity development. *Environmental Education Research, 22*(7), p. 1032. https://www.tandfonline. com/doi/abs/10.1080/13504622.2015.1072136 Reprinted with permission of Taylor and Francis.

of comfort and security in the natural world while at the same time ensuring that they are aware of and can successfully navigate nature's inherent dangers. Teaching children how to negotiate aspects of the environment that may pose threats and cause anxiety or fear can foster an inner sense of *Trust in Nature*. For example, while children might initially be frightened when a spider is crawling along a log beside them, modeling to the children how to reposition themselves calmly and contently will impart dispositions and self-competencies that are essential for negotiating similar conflicts in different situations and contexts. This example could also be considered in children's navigation between *Environmental Competency* and *Environmental Disdain*. Contempt or disdain is often caused by a reaction or fear of the unknown. Thus, supporting children in learning how to identify various types of spiders that may inhabit a particular environment instills a sense of *Environmental Competency*, providing both knowledge and feelings of confidence in navigating an environmental dilemma. In this way, I argue that negotiating tensions inherent in the natural world plays an essential role in healthy EID, strengthening feelings of self-confidence and connections with the natural world.

The Role of Emotions in EID[3]

Throughout this opening chapter I have alluded to the emotional dimensions of children's EID. Indeed, a child's affective states, or feelings, determine how he or she interacts with and relates to the natural world. While nature has long been recognized as a setting that stimulates children's curiosity and wonder (Carson, 1956; Wilson, 1997), that is their positive emotional encounters, less attention has been given to the negative emotions (e.g., fear and anxiety) that signifies their experience with their environment. Additionally, the literature scarcely references the role caregivers and educators play in supporting children in overcoming the emotional tensions that may be experienced in nature.

Emotional development occurs rapidly during the early years of life. From the expression of a few basic emotions (love, joy, anger, sadness, and fear) in infancy, children develop and exhibit a wide range of emotions including frustration, worry, pride, and guilt by the first few years of life (Boyer, 2014). Boyer (2014) characterized emotions as having similar properties as a reaction: "It often has an identifiable cause or stimulus; it is usually brief, spasmodic, intense experience of short duration; and the person is typically much aware of it" (p. 11). Emotions can be distinguished from a mood, where "a mood tends to be more subtle, longer lasting, less intense, more in the background, a frame of mind, casting a positive or negative light over experiences" (Boyer, 2014, p. 11). Both emotions and moods, however, influence children's engagement towards and are influenced by their environments. Consistent and repeated adverse emotional experiences in an environment may cast a negative mood to be associated with that setting thus, both moods and emotions influence an individual's EID.

As their repertoire of emotions expands, young children begin to identify and recognize their own feelings and the feelings of others. They attach labels, or names, to identify basic emotions, pairing specific emotions with certain facial expressions or behaviors (e.g.. that girl feels sad because she is crying) (Boyer, 2014). Empathy also emerges during early childhood and is generally applied towards familiar people or objects. Empathy, or the ability to feel what someone else is feeling, promotes pro-social and helping behaviors as well as moral decision-making (McDevitt & Ormrod, 2013).

Empathy also informs the way children relate with nature. Chawla (1998) noted two ways in which a child may express empathy towards an environment. In the first, children might perceive an environment as having no intrinsic feelings, thus, they would project their own feelings onto that environment. This is closely related to anthropomorphism, which involves attributing human qualities to non-human entities. By attributing human-like qualities to, say, a tree or a beetle, children express "feelings of empathy for the object that permit it to be regarded as something worthy of moral consideration" (Gebhard et al., 2003, p. 92). The second expression of empathy towards an environment involves recognizing ecosystems as "living wholes with intrinsic intelligence, feelings, needs, or rights" (Chawla, 1998, p. 12). Similarly, Hungerford and Volk (1990) referred to environmental sensitivity as "an empathetic perspective towards the environment" (p. 11). However, Chawla (1998) argued that "environmental sensitivity is not empathy, but instead a predisposition to take an interest in learning about the environment, feeling concern for it, and acting to conserve it on the basis of formative experiences" (p. 19). In other words, while empathetic feelings may inform environmental sensitivity, environmental sensitivity also involves knowledge about the environment, which together promotes environmental responsibility or action (Metzger & McEwen, 1999).

Another important socio-emotional skill gained during the early years is self-regulation, or the "process of directing and controlling one's personal actions and emotions" (McDevitt & Ormrod, 2013, p. 79). Emotional regulation refers specifically to the strategies children development to manage their affective states (McDevitt & Ormrod, 2013). Children learn to regulate their actions and behaviors resulting from both positive and negative emotions. For example, when young children feel angry, they may react by physically hitting another child. With guidance from adults, children eventually learn how to constructively express their anger verbally instead of physically lashing out (Cole, Armstrong, & Pemberton, 2010). Young children learn cultural norms and standards for emotional expression, including what behaviors and actions are acceptable and unacceptable in various settings (Boyer, 2014; Morelli & Rothbaum, 2007), including home, school, *and* the natural environment. However, little has been written about children's emotional regulation in nature. While some have discussed nature as a restorative setting for processing and regulating one's feelings (Kaplan, 1995; Korpela, Hartig, Kaiser, & Fuhrer, 2001), none have discussed how young children learn to regulate their emotions and behaviors in response to natural stimuli.

Emotions and Learning

Emotions are fundamental to learning. Neuroscience shows that the accomplishment of a learning task is dependent on three brain networks (recognition, strategic, and affective) working together (Hinton, Miyamoto, & Della-Chiesa, 2008). The recognition network "receives sensory information from the environment and transforms it into knowledge;" the strategic network is responsible for "planning and coordinating goal-oriented actions;" and the affective network is "involved in the emotional dimensions of learning such as interest, motivation and stress" (Hinton et al., 2008, p. 91). Fear and stress hinders the recognition and strategic network of the brain, which activates learning.

While negative emotions can be detrimental to learning, positive emotions can drive a person's motivation to engage in learning. Situations influence emotions and emotions direct actions thus situations associated with positive emotions are approachable and desirable (Boyer, 2015). A primary goal then is to encourage intrinsic motivation, or children's innate desire to explore and learn. Educators can increase intrinsic motivation through fostering children's sense of self-efficacy, competency, and autonomy and by relating learning to matters that are important to them (Hinton et al., 2008; McDevitt & Ormrod, 2013).

Emotions and Environmental Education

Research in environmental education has also examined how emotions influence environmental behavior (Camri, Arnon, & Orion, 2015; Zeyer & Kelsey, 2013). These studies revealed that although students possessed knowledge about the environment, knowledge (objective or subjective understanding) is not enough to stimulate environmental behavior; in turn, environment knowledge *must* be mediated through feelings or emotions one holds towards their environment (Camri et al., 2015). Specifically, Zeyer and Kelsey (2013) found that although youth can readily recite the facts regarding the global state of the environment, their response to these concerns portrays a "pessimistic mood," "motive of guilt," and "lack of feeling of control" (p. 207). Subsequently, environmental education approaches must take note of the affective domain and seek to address the tensions that have rooted environmental hope or environmental despair (Green, 2016; Hicks, 2014; Kelsey & O'Brien, 2011).

Two decades ago, Sobel (1996) argued that environmental education approaches were placing too much emphasis on teaching about environmental problems, resulting in children developing a sense of helplessness or fear of nature. Global issues such as oil spills and now global climate change were being placed on children's shoulders much too early; instead Sobel (1996) argued "that children [should] have the opportunity to bond with the natural world, to learn to love it and feel comfortable in it, before being asked to heal its wounds" (p. 13). In line with Sobel's (1996) argument, research on formative childhood experiences shows that early emotional connection to nature plays a significant role in shaping environmental values and behaviors (Chawla, 1998; Louv, 2008; Well & Lekies, 2006). Thus, connecting or reconnecting children to nature has become one of the primary goals of environmental education.

Indeed early play opportunities in nature provide many physical (Fjørtoft, 2001), as well as psychological and social benefits including stimulating fantasy and imaginary activities, promoting discovery and exploration, encouraging risk-taking and the development of environmental competencies, and fostering a sense of autonomy (Chawla & Rivkin, 2014; Dowdell, Gray, & Malone, 2011; Green, 2011; 2013; 2015a). Little research, however, has specifically focused on the emotional dimensions of young children's encounters with nature, although these can somewhat be inferred by the literature. It is widely recognized that positive experiences in nature invokes all the senses, promotes a sense of wonder, instills aesthetic sensitivity, builds children's self-confidence and a healthy self-concept, and promotes nurturing and caring behaviors (Carson, 1956; Elliott, 2010; Louv, 2008; Wilson, 2012). Negative or lack of experiences in nature may result in depression and anxiety or fear and insecurity (Bixler et al., 1994; Louv, 2008). Thus, my book aims to provide further understanding of children's emotional encounters in nature, with a particular focus on how affective states influence a child's EID.

Conclusion

In this chapter, a new model for Environmental Identity Development (EID) is presented, which considers young children's developing self-cognitions (i.e. affective states, values, beliefs, and preferences) in relation to the natural world. While environmental scholars have long recognized the significance of children's early experiences with nature, EID provides a framework to understand *how* young children's environmental identity progresses during the critical years of early childhood. The EID model draws from contemporary sociological understandings of childhood that recognize young children's agency (Corsaro, 2015; James & Prout, 1990) as well as psychosocial theories of identity development (Erikson, 1950, 1972, 1980). Additionally, the EID model is supported by current literature in early childhood environmental education and education for sustainability, with a particular emphasis on studies that consider the experiences, perspectives, and voices of young children.

The proceeding chapters in this book further explore the various stages of the EID model. Chapter Two explains the foundational stage of *Trust in Nature* vs. *Mistrust in Nature*, asserting that feelings of assurance and comfort in nature vs. feelings of anxiety and discomfort are fundamental to one's ability to progress along the continuum of healthy EID. From this foundational aspect of trust, Chapter Three examines the importance of *Spatial Autonomy*, where children venture out from caregivers to discover their own sense of place both independently and collectively with other children. Contrary to gaining *Spatial Autonomy* are feelings of doubt or *Environmental Shame* associated with nature experiences, occurring when children are discouraged or prohibited from exploring a self-concept in nature. Chapter Four discusses the development of *Environmental Competency*. Stemming from continued positive experiences in nature, children develop skills and confidence to use the environment to carry out personal and social goals and enrich their experiences. Once again, lack of opportunities or discouragement, could lead to

Environmental Disdain or cognitions of contempt that separate children from the natural world. Chapter Five describes how a strong sense of *Environmental Competency* drives children to exercise agency in participating in *Environmental Action*, applying values, care, and ethics to create a more sustainable future whereas children who lack in their development of *Environmental Competency* may form disregard for nature, displaying dispositions and behaviors that lead to *Environmental Harm*. In addition, each chapter considers sociocultural and geographical dimensions and examples of how particular spatial and temporal contexts may influence children's environmental identity.

Chapter Six considers implications and future directions for an integrated EID model, including discussion of methodological and ethical considerations for childhood place and environmental education research. The EID model can serve as a theoretical framework for research in environmental education and education for sustainability, providing a lens to consider the way in which an individual progresses in developing his/her environmental identity. However, in applying EID to research, scholars must consider participatory methodological approaches in research *with* or *by* children as opposed to research conducted *on* children (see Barratt-Hacking, Cutter-Mackenzie, & Barratt, 2013; Green, 2015b), as the EID model recognizes children's autonomy and agency as essential in empowering children to work for a sustainable future.

Finally, Chapter Seven broadens understanding of EID by exploring the diversity of children's experiences. Pathways of EID differ depending on cultural and geographical locations, familial influences, and children's own interests. This chapter extends the conversation by including essays from three Alaskan educators. Each author reflects on EID in different Alaskan setting as well as other parts of the world. Additionally, this final chapter extends educational applications of EID by considering a transnational project with youth. Children's environmental identity develops in diverse ways and is highly influenced by family, sociocultural, and geographical contexts. Therefore, in considering support strategies, an educator should start by: (1) carefully considering the needs of individual children and the collective group as a whole and (2) identifying appropriate entry points for supporting individual needs and the group as a whole.

Finally, my own personal reflections are included in each chapter to further explore how EID is constructed in various circumstances and situations. In this way, the reader is invited to actively reflect on his or her own EID, against the backdrop of personal life experiences.

Notes

1. Copyright 2016 from "A Twilight Dance in the Snow: Thinking about Environmental Identity Development" by Carie Green and Michael Brody in K. Winograd (Ed.), *Education in times of Environmental Crises: Teaching children to be agents of change* (pp. 23–25). Reproduced by permission of Taylor and Francis Group, LLC, a division of Informa plc. This permission does not cover any third party copyrighted work, which may appear in the material requested.

2. The introduction to EID theory was first proposed in Green, C., Kalvaitis, D., & Worster, A. (2016). Recontextualizing psychosocial development in young children: A model of environmental identity development. *Environmental Education Research, 22*(7), 1025–1048. https://www.tandfonline.com/doi/abs/10.1080/13504622.2015.1072136 Content of the article reproduced in this chapter with permission from Taylor and Francis.

3. The Canadian Journal of Environmental Education has given permission to reprint a portion of the following article in this chapter: C. Green (2016). Monsters or good guys: The mediating role of emotions in transforming a young child's encounter with nature. Canadian Journal of Environmental Education, *21*, 125–144.

References

Ainsworth, M. D. S., Blehar, M., Waters, E., & Wall, S. (1978). *Patterns of attachment: A psychological study of the strange situation*. Hillside, NJ: Erlbaum.

Barratt Hacking, E., Cutter-Mackenzie, A., & Barratt, R. (2013). Children as active researchers: The potential of environmental education research involving children. In R. B. Stevenson, M. Brody, J. Dillon, & A. E. J. Wals (Eds.), *International handbook of research on environmental education* (pp. 438–458). New York, NY: Routledge.

Barratt, R., Barratt-Hacking, E., & Black, P. (2014). Innovative approaches to early childhood education for sustainability in England. In J. Davis & S. Elliott (Eds.), *Research in early childhood education for sustainability: International perspectives and provocations* (pp. 225–247). New York, NY: Routledge.

Bixler, R. D., Carlisle, C. L., Hammitt, W. E., & Floyd, M. F. (1994). Observed fears and discomforts among urban students on field trips to wildland areas. *The Journal of Environmental Education, 26*(1), 24–33.

Bixler, R. D., & Floyd, M. F. (1997). Nature is scary, disgusting, and uncomfortable. *Environment and Behavior, 29*(4), 443–467.

Blatt, E. (2014). Uncovering students' environmental identity: An exploration of activities in an environmental science course. *The Journal of Environmental Education, 45*(3), 194–216.

Boyer, G. H. (2014). How might emotions affect learning. In S. A. Christianson (Ed.), *The handbook of emotion and memory: Research and theory* (pp. 3–32). New York, NY: Psychology Press.

Brügger, A., Kaiser, F. G., & Roczen, N. (2011). One for all? *European Psychologist. 16*(4), 324–333.

Camri, N., Arnon, S., & Orion, N. (2015). Transforming environmental knowledge into behavior: The mediating role of environmental emotions. *The Journal of Environmental Education, 46*(3), 183–201.

Carson, R. (1956). *The sense of wonder.* New York, NY: Harper & Row.

Chawla, L. (1998). *Significant life experiences revisited: A review of research on sources of environmental sensitivity. The Journal of Environmental Education, 29*(3), 11–21.

Chawla, L., & Rivkin, M. (2014). Early childhood education for sustainability in the United States. In J. Davis & S. Elliott (Eds.), *Research in early childhood education for sustainability: International perspectives and provocations* (pp. 248–265). New York, NY: Routledge.

Clayton, S. (2003). Environmental identity: A conceptual and an operational definition. In S. Clayton and S. Opotow (Eds.), *Identity and the natural environment* (pp. 45–65). Cambridge, MA: MIT Press.

Clayton, S., Fraser, J., & Burgess, C. (2011). The role of zoos in fostering environmental identity. *Ecopsychology 3*(2), 87–96.

Clayton, S., & Opotow, S. (2003). *Identity and the natural environment.* Cambridge, MA: MIT Press.

Cole, P. M., Armstrong, L. M., & Pemberton, C. K. (2010). The role of language in the development of emotion regulation. In S. D. Calkins, & M. A. Bell (Eds.), *Child development at the intersection of emotion and cognition* (pp. 59–77). Washington, DC: American Psychological Association.

Corsaro, W. A. (2015). *The sociology of childhood* (4th ed.). Thousand Oaks, CA: Sage.

Davis, J. & Elliott, S. (2014). *Research in early childhood education for sustainability: International perspectives and provocations.* New York, NY: Routledge.

Dowdell, K., Gray, T., & Malone, K. (2011). Nature and its influence on children's outdoor play. *Australian Journal of Outdoor Education, 15*(2), 24–35.

Dunkel, C. S., & Sefcek, J. A. (2009). Eriksonian lifespan theory and life history theory: An integration using the example of identity formation. *Review of General Psychology, 13*(1), 13–23.

Elliott, S. (2010). Children in the natural world. In J. M. Davis (Ed.), *Young children and the environment: Early education for sustainability* (pp. 43–75). New York, NY: Cambridge University Press.

Erikson, E. H. (1950). *Childhood and society* (1st ed.). New York: Norton & Company.

Erikson, E. H. (1972). Eight stages of man. In C. S. Lavatelli & F. Stendler (Eds.), *Readings in child behavior and child development* (pp. 19–30). San Diego, CA: Harcourt Brace Jovanovich.

Erikson, E. H. (1980). *Identity and the life cycle.* New York: Norton & Company.

Evans, E., Ching, C. C., & Ballard. H. L. (2012). Volunteer guides in nature reserves: Exploring environmental educators' perceptions of teaching, learning, place and self. *Environmental Education Research, 18*(3), 391–402.

Fjørtoft, I. (2001). The natural environment as a playground for children: The impact of out-door play activities in pre-primary school children. *Early Childhood Education Journal, 29*(2), 111–117.

Gebhard, U., Nevers, P., & Billmann-Mahecha, E. (2003). Moralizing trees: Anthropomorphism and identity in children's relationship to nature. In S. Clayton & S. Opotow (Eds.), *Identity and the natural environment* (pp. 91–112). Cambridge, MA: MIT Press.

Green, C. (2011). A place of my own: Exploring preschool children's special places in the home environment. *Children, Youth, and Environments, 21*(2), 118–144.

Green, C. (2013). A sense of autonomy in young children's special places. *International Journal of Early Childhood Environmental Education, 1*(1), 8–33.

Green, C. (2015a) "Because we like to": Young children's experiences hiding in their home environment. *Early Childhood Education Journal, 43*(4), 327–336.

Green, C. (2015b). Towards young children as active researchers: A critical review of the methodologies and methods in early childhood environmental education research. *Journal of Environmental Education, 46*(4), 207–229.

Green, C. (2016). Sensory tours as a method for engaging children as active researchers: Exploring the use of wearable cameras in early childhood research. *International Journal of Early Childhood, 48*(3), 277–294.

Green, C., Kalvaitis, D., & Worster, A. (2016). Recontextualizing psychosocial development in young children: A model of environmental identity development. *Environmental Education Research, 22*(7), 1025–1048.

Hägglund, S., & Johansson, E. M. (2014). Belonging, value conflicts, and children's rights in learning for sustainability in early childhood. In J. Davis & S. Elliott (Eds.), *Research in early childhood education for sustainability: International perspectives and provocations* (pp. 38–48). New York, NY: Routledge.

Hart, R. (1979). *Children's experience of place.* New York, NY: Irvington.

Hicks, D. (2014). A geography of hope. *Geography, 99*(1), 5–12.

Hinds, J., & Sparks, P. (2009). Investigating environmental identity, well-being, and meaning. *Ecopsychology, 1*(4), 181–186.

Hinton, C., Miyamoto, K., & Della-Chiesa, B. (2008). Brain research, learning and emotions: Implication for education research, policy, and practice. *European Journal of Education, 43*(1), 87–103.

Holmes, S. J. (2003). Some lives and some theories. In S. Clayton and S. Opotow (Eds.), *Identity and the natural environment* (pp. 25–42). Cambridge, MA: MIT Press.

Hungerford, H. R., & Volk, T. L. (1990). Changing learner behavior through environmental education. *The Journal of Environmental Education, 21*(3), 8–21.

James, A., & Prout, A. (1990). *Constructing and reconstructing childhood: Contemporary issues in the sociological study of childhood.* Bristol, PA: Taylor and Francis.

Kals, E., & Ittner, H. (2003). Children's environmental identity: Indicators and behavioral impacts. In S. Clayton and S. Opotow (Eds.), *Identity and the natural environment* (pp. 135–154). Cambridge, MA: MIT Press.

Kaplan, S. (1995). The restorative benefits of nature: Toward an integrative framework. *Journal of Environmental Psychology, 15*(3), 169–182.

Kelsey, E., & O'Brien, C. (2011). Sustainable happiness. *Green Teacher, 93*, 3–7.

Korpela, K. M., Hartig, T., Kaiser, F. G., & Fuhrer, U. (2001). Restorative experience and self-regulation in favorite places. *Environment and Behavior, 33*(4), 572–589.

Lee, N. (2001). *Childhood and society: Growing up in an age of uncertainty.* Buckingham: Open University Press.

Lerner, R. M. (2002). *Concepts and theories of human development* (3rd ed.). Mahwah, NJ: Erlbaum.

Louv, R. (2008). *Last child in the woods: Saving our children from nature-deficit disorder.* (expanded version.). Chapel Hill, NC: Algonquin.

Lowan-Trudeau, G. (2013). Indigenous environmental education research in North America: A brief review. In R. B. Stevenson, M. Brody, J. Dillon, & A. E. J. Wals (Eds.), *International Handbook of research on environmental education* (pp. 404–408). New York, NY: Routledge.

Lucas, A. M. (1979). *Environment and environmental education: Conceptual issues and curriculum implications.* Melbourne, Australia: Australia International Press and Publications.

Manzo, L. C., & Devine-Wright, P. (2014). *Place attachment: Advances in theories, methods, and applications.* New York, NY: Routledge.

McDevitt, T. M., & Ormrod, J. E. (2013). *Child development and education* (5th ed.). Upper Saddle River, NJ: Pearson.

Metzger, T., & McEwen, D. (1999). Measurement of environmental sensitivity. *The Journal of Environmental Education, 30*(4), 38–39.

Morelli, G. A., & Rothbaum, F. (2007). Situating the child in context: Attachment relationships and self-regulation in different cultures. In S. Kitayama & D. Cohen (Eds.), *Handbook of cultural psychology* (pp. 500–527). New York, NY: Guilford Press.

Müller, K. M., Kals, E., & Pansa, R. (2009). Adolescents' emotional affinity toward nature: A cross-societal study. *Journal of Developmental Processes, 4*(1), 59–69.

Palmer, J. A., & Suggate, J. (1996). Influences and experiences affecting the pro-environmental behaviour of educators. *Environmental Education Research, 2*(1), 109–121.

Payne, P. (2001). Identity and environmental education. *Environmental Education Research, 7*(1): 67–88.

Prévot, A. C., Clayton, S., & Mathevet, R. (2018). The relationship of childhood upbringing and university degree program to environmental identity: Experience in nature matters. *Environmental Education Research, 24*(2), 263–279.

Proshansky, H. M., & Fabian, A. K. (1987). The development of place identity in the child. In C. S. Weinstein & T. G. David (Eds.), *Spaces for children: The built environment and child development* (pp. 21–40). New York, NY: Plenum Press.

Qvortrup, J., Corsaro, W. A., & Honig, M.-S. (2009). *The Palgrave handbook of childhood studies.* London, UK: Palgrave Macmillan.

Rasmussen, S. (2009). Commentary: Opening up perspectives on autonomy and relatedness in parent–children dynamics: Anthropological insights. *Culture and Psychology, 15*(4): 433–449.

Ritchie, J. (2014). Learning from the wisdom of elders. In J. Davis & S. Elliott (Eds.), *Research in early childhood education for sustainability: International perspectives and provocations* (pp. 49–60). New York, NY: Routledge.

Scannell, L., & Gifford, R. (2010). Defining place attachment: A tripartite organizing framework. *Journal of Environmental Psychology, 30*(1), 1–10.

Seamon, D. (2014). Place attachment and phenomenology: The synergistic dynamism of place. In L. C. Manzo & P. Devine-Wright (Eds.), *Place attachment: Advances in theories, methods, and applications* (pp. 11–22). New York, NY: Routledge.

Sebba, R. (1991). The landscapes of childhood: The reflection of childhood's environment in adult memories and in children's attitudes. *Environment and Behavior, 23*(4), 395–422.

Sobel, D. (1996). *Beyond ecophobia: Reclaiming the heart of nature education.* Great Barrington, MA: Orion.

Stapleton, S. (2015). Environmental identity development through social interactions, action, and recognition. *The Journal of Environmental Education, 46*(2), 94–113.

Stets, J. E., & Biga, C. F. (2003). Bringing identity theory into environmental sociology. *Sociological Theory, 21*(4), 398–423.

Thomashow, M. (1995). *Ecological identity: Becoming a reflective environmentalist.* Cambridge, MA: MIT Press.

Tugurian, L. P., & Carrier, S. J. (2017). Children's environmental identity and the elementary science classroom. *The Journal of Environmental Education, 48*(3), 143–153.

United Nations. (1989). *Convention for the rights of the child.* New York, NY: Author.

United Nations. (2005). *Convention on the rights of the child: General Comment No. 7. Implementing child rights in early childhood.* Geneva, Switzerland: Author.

Wells, N. & Lekies, K. (2006). Nature and the life course. *Children, Youth and Environments, 16*(1), 1–24.

Williams, C. C., & Chawla, L. (2016). Environmental identity formation in nonformal environmental education programs. *Environmental Education Research, 22*(7), 978–1001.

Wilson, R. A. (1996). The development of the ecological self. *Early Childhood Education Journal, 24*(2), 121–123.

Wilson, R. A. (1997). The wonders of nature: Honoring children's ways of knowing. *Early Childhood News, 9*(2), 6–9.

Wilson, R. A. (2012). *Nature and young children: Encouraging creative play and learning in natural environments.* New York, NY: Routledge.

Zeyer, A. & Kelsey, E. (2013). Environmental education in a cultural context. In R. B. Stevenson, M. Brody, J. Dillon, & A. E. J. Wals (Eds.), *International handbook of research on environmental education* (pp. 206–212). New York, NY: Routledge.

· 2 ·

TRUST IN NATURE VS. MISTRUST
IN NATURE

A Sense of Trust

Inscribed in my memory is an image of myself hanging on a rocky ledge of a sharp cliff over the turquoise ocean. I had not planned on being there but somehow that is where I had ended up. I had trekked across grazing sheep fields to find what seemed to be a trail across a red sandy beach that suddenly ended and as I balanced on the sandstone cliff, I realized that the only direction to head was up. I remember slowly taking off my heavy traveling backpack, lifting it over my head, and with all of my strength nudging it onto the ledge above me. Now that the backpack had made it safely, I had no choice but to make it too. But how? If I looked down for too long at the bellowing waves that swelled and whipped against the rocks below me, I might become defeated by the reality of my situation. Yet if I stepped too far out of that reality, I might not be able to muster the strength to make it over the cliff. What was left was to rely on my faith and a deep sense of trust that I could navigate this outer environmental dilemma and muster the strength to pull my own body up and onto the rock above me.

This same deep sense of *Trust in Nature* has been my "go to" in other really hairy situations. I remember once hiking to the top of Mt. Elden in Flagstaff,

Arizona, taking in the view at the top, and realizing that a dark storm was soon to hit. I took off running down the mountain as fast as I could but before I made it a quarter of the way down, hail was blazing at my back. I had to trust and find the strength to endure and seek refuge somewhere and before I knew it I found a small dwelling under a rock. Another time I was trekking in New Zealand and somehow lost track of the main trail and was unable to find the hut where I had planned to sleep that night. Trust enabled me to roll out my sleeping bag and sleep in the bush under a Tōtara tree. This enduring sense of *Trust* in *Nature* has become a part of my environmental identity, which provides me confidence to work through the many challenges that I have faced in various environments. It is the basis upon which other attributes of my environmental identity have been shaped and formed.

What is *Trust in Nature?* That is the subject of this chapter. But before unraveling the meaning of trust in an environmental context, it is important to understand that such trust does not come without the potential for mistrust, the misguided belief that trust cannot be had and will not prevail in any given environmental dilemma or situation. Trust is defined as "assured reliance on the character, ability, strength, or truth of someone or something" and someone or something in which "confidence is placed" (Merriam-Webster, 2018). In comparison, mistrust is a "lack of confidence" in something or someone (Merriam-Webster, 2018). Synonyms of mistrust include "uncertainty, doubt, skepticism, and suspicion" (Merriam-Webster, 2018). Trust in the context of this book will be discussed as a social element of an individual's constructed relationship with nature and all that exists. Trust can be formed individually or with someone or something else, but it is a prevailing sense in the good that sustains our encounters with nature.

According to Erikson (1980), trust is the fundamental social dimension of all positive relationships established with another, developed during the early stages of life between an infant and their primary caregiver. In this chapter, I extend the context of trust to include an individual's growing relationship with the natural environment. Figure 2.1 includes basic attributes that relate to a sense of *Trust in Nature* vs. *Mistrust in Nature*. These inner attributes of trust are encountered through experiences in nature and strengthened when one faces certain outer environmental dilemmas.

I argue that *Trust in Nature* is not established overnight but develops over time and is established through meaningful encounters in nature. It is revisited constantly throughout one's life and strengthened through each and every experience that one has in the natural world. *Trust in Nature* is the inner

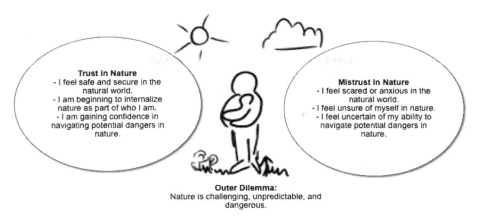

Figure 2.1. Trust in Nature vs. Mistrust in Nature © Carie Green 2018.

sense of peace, security, and safety that one feels in nature. It is the image of Mother Earth holding an innocent life close as depicted in the picture above. This image can also refer to the security and comfort between a caregiver and child during initial encounters in nature. In the early years, a caregiver's role in ensuring that a child feels safe, comfortable, and secure during nature experiences is immeasurable. Through continuous exposure, a child will begin to internalize nature as an essential part of who they are. And through support from others and eventually through one's own inner conviction an individual will prevail with confidence in their abilities to navigate challenges and perils that are inherent in the natural world.

In the same way, *Mistrust in Nature* is also fostered and nurtured through experiences with the natural world. *Mistrust in Nature* might be caused by fear conditioning (Phillips & LeDoux, 1992), resulting from a particular adverse experience in nature that imprinted strong feelings of fear and anxiety. Anxiety and fear, or a sense of *Mistrust in Nature*, might also result from observing others' reactions to environmental stimuli. For example, a student in my child development class once told me about her son's fear of insects. When I asked her if she was afraid of insects, she said yes. As a result, it seemed that her son had learned this behavior from his parent. *Mistrust in Nature* might also result from less frequent or sporadic encounters with nature where trust does not have the opportunity to take root. Through such adverse experiences with nature, a child might over time begin to feel unsure of who they are while in nature. Furthermore, a child might fail to develop a lack of confidence in their ability to navigate potential dangers and challenges posed during nature experiences.

To begin, we will explore the psychological development of *Trust in Nature* by revisiting Erikson's (1950, 1972, 1980) theory of psychosocial development and Bowlby's (1980) attachment theory. Then we examine place attachment theory, specifically focusing on human relations with natural places. Next, we will build on the emotional aspects of psychosocial development and attachment in conceptualizing *Trust in Nature* as the foundation of EID. Then we will discuss a caregiver's role in providing young children with secure and comfortable encounters in nature. Throughout this chapter, I will provide examples of how *Trust in Nature* emerges distinctly in each child and is dependent upon personality, social interactions, culture, and through guidance and education. Lastly, we will look at future directions for research and application in understanding how *Trust in Nature* emerges in childhood and throughout life.

Revisiting Psychosocial Development and Attachment Theory

When children are born they are fully dependent on caregivers for all their physiological and psychological needs. According to Erikson's (1950, 1972, 1980) psychosocial development theory, during the first stage of *Trust vs. Mistrust*, infants learn to trust their primary caregiver(s) to provide for their physical needs such as satisfying hunger and changing an uncomfortable diaper as well as their socio-emotional needs such as receiving comfort and affection. When care from their caregiver(s) is consistent, children establish a sense of security and trust. Trust, according to Erikson (1980) is "commonly implied in reasonable trustfulness as far as others are concerned and a simple sense of trustworthiness as far as oneself is concerned" (p. 57). In other words, trust is associated with how one relates to others and how one relates to oneself. Trust, therefore, is essential for healthy social interactions and forming and maintaining healthy relationships later in life.

Similarly, attachment theory emphasizes the importance of adult proximity and caregiving responses to infant behaviors (Bowlby, 1980). Moving beyond behavioristic philosophies, attachment theory is premised on the emotional bonding that occurs between children and their caregivers/attachment figure. Bowlby (1980) argued that long-term emotional bonds are a basic part of human nature. Such trusting bonds enable feelings of security and well being. In contrast, when care and affection from an attachment figure is inconsistent, infants develop feelings of anxiety and distress, which over time may result in a sense of mistrust. Furthermore, enduring feelings of mistrust may cause an individual to perceive others and the world as unsafe and unpredictable.

Thus, an infant's daily interactions and experiences with their primary caregiver/ attachment figure shape unconscious psychological structures, comprising implicit memories that form an internal working model (Bowlby, 1980). This internal working model becomes the template for structuring all subsequent social relationships. Contemporary research of human attachment theory also links early attachments to healthy brain development where secure attachments lead to the development of a large number of synaptic connections. An unhealthy attachment relationship, however, can have a negative impact on synaptic connectivity and psychological functioning (Greenspan, 1999).

In this way, early attachments influence identity development in that the "psychological structure of the self emerges from the inter-subjective context of the attachment relationship" (Morgan, 2010, p. 13). Through consistently responding to the biological needs expressed by an infant, a caregiver helps an infant develop an internal working model of trust and emotional regulation: "The process of emotional regulation that emerges in the attachment relationship plays a major role in establishing internal coherence in the infant" (Morgan, 2010, p. 13). Thus, a positive sense of self emerges through continuous secure and responsive interactions between an infant and their primary caregiver laying the foundation upon which future aspects of identity development will eventually be built.

Place Attachment and Trust in Nature

Attachment theory has been adapted in the field of environmental psychology to explain how humans form attachments to natural and human-built environments. Particularly, place attachment theory refers to the positive bond that a person develops towards a place (Low & Altman, 1992). Sometimes referred to as a sense of place (Hay, 1998), it is primarily informed by an individual's emotional attachment towards a place (Chawla, 1992; Devine-Wright & Clayton, 2010; Low & Altman, 1992), but it may also include knowledge and beliefs as well as behaviors and actions towards a place (Proshansky, Fabian, & Kaminoff, 1983; Scannell & Gifford, 2010). While place attachments fluctuate and change over time (Hay, 1998), enduring attachments to places often develop during childhood (Chawla, 1992; Morgan, 2010).

Place attachments are concerned with specific locales (e.g., a rock by a river, a barn loft), which can range in scale from large to small. Although place attachments may refer to places in nature, it may also include an individual's relationship to human-built places and does not necessarily focus on identity formation in the natural world. That is, place attachments are specific to each individual and may or may not include natural places. While the topic of this book is EID, focusing on how an individual's sense of self develops in relation to nature, the place attachment literature helps us to understand the important role of emotions in the development of human relationships with their environment. Secure attachments to places emerge through recurring positive experiences (Morgan, 2010). What can also be infered from the literature, although not specifically stated, is the importance of trust as the basis for the development of one's relationship with the natural world.

Trust in Nature as the Foundation for EID

Extending Erikson's first stage of identity development (*Trust vs. Mistrust*), I propose that in the same way that children form secure and trusting attachments with caregivers, it is also important for children to develop a trusting bond with the natural world. This trusting bond, established through early, shared experiences in nature, is essential to positive EID. As noted, initially, these shared experiences often involve a primary caregiver. However, as a child grows such experiences may also involve other adults as well as siblings and peers. If these initial shared experiences with nature are positive and consistent, a trusting relationship with a caregiver and nature will be strengthened and the seed of potential positive identification with the environment will be formed (Wilson, 1996). If, however, experiences in nature are infrequent and characterized by anxious encounters, then children may develop mistrust and insecure bonds with the natural world, which has been shown to have negative implications for an individual's environmental identity later in life (Bixler, Carlisle, Hammitt, & Floyd, 1994; Bixler & Floyd, 1997).

Building on place attachment theory, children are most likely to form attachments to specific familiar places in nature that are associated with strong emotional ties. And like the emotional bonding that occurs between an infant and their attachment figure, a young child's interactions and experiences in their environment will shape unconscious psychological structures, forming an internal working model that influences the way they relate and identify with the natural world. And like human or place attachments, the associations that children make with the natural world may be good or bad, positive or negative, or a mixture of both therefore it is essential that caregivers, family, and peers help children develop trusting and secure associations that contribute to their sense of well being.

Foundational aspects of an individual's environmental identity develop in the first years of life. As Chawla (2008) explains, "even before they learn to talk, their disposition to engage with the world and to make their mark – a disposition essential for community participation – is rooted in infancy" (p. 99). Research is beginning to show the many positive effects of being outdoors even for the very young (Wells, 2000; Wilson, 1995). Allowing direct nature experiences that create sensations of beauty and mystery provide ripe opportunities for EID.

Just as children who form secure attachments with their caregivers are more confident, competent, and often display better social interactions with peers later in life (Groh, Fearson, Bakersmans-Kranenburd, Van IJsendoorn, Steele, & Roisman, 2014), children who develop trusting relationships with nature early on will be eager to explore and act for the environment in future stages. In this way, engagement *in/with* nature is a dynamic and reciprocal process where even young children respond positively to self-directed interactions with their natural environment (Heft & Chawla, 2006). If children lack experiences *in* nature they will bond less deeply and possibly develop mistrust in that environment. Learning *from* nature at this stage occurs instinctually as young children soak up sensory information and grow neurological connections. Immersed in nature, a child attempts to make sense of all they see, hear, touch, taste, and smell. In this way, a child is forming essential aspects of their environmental identity.

Establishing Trust in Infancy: A Caregiver's Role

During the first twelve months of life, infants are physiologically and psychologically dependent on caregivers for their basic needs. The caregiver's role then is to provide consistent care and developmental experiences rich in sensory and language stimuli. Erikson (1980) described this as the "*incorporative stage*" in which an infant is "receptive to what is being offered" (p. 59). Erikson (1980) goes on to explain infancy as a vulnerable and sensitive period. Thus, it is important that caregivers "deliver to their senses stimuli as well as food in the proper intensity and at the right time; otherwise their willingness to accept may change abruptly into diffuse defense – or into lethargy" (p. 59). In other words, a caregiver should see to it that an infant's basic needs are met before exposing them to novel environmental stimuli. Furthermore, Chawla and Rivkin (2014) argue that children learn how to respond to the environment by way of "joint attention and social referencing" (p. 252) between caregiver and child. Recall the example of a child who learned to fear insects from his mother.

Although nature provides an optimal environment for sensory stimulation, infants in many Western cultures are often confined to cribs, car seats, and walkers even by caregivers with the best of intentions. Caregivers need also consider the importance of providing young children with positive and frequent encounters in nature as such experience play an essential role in fostering children's bonds with nature. However, in bringing young children outdoors it is important to ensure they are dressed appropriately (e.g., protected from the sun and kept warm when its cold) because comfort and security are critical components in building trusting relations with both caregivers and the environment. For example, caregivers may set up daily "tummy play" for infants in an outdoor setting. This will allow infants opportunities to soak up the sights, sounds, smells, and textures only available through such close contact with nature. Caregivers can use language to talk about what is observed close by and point out interesting colors and movement. They should remain close, making time for bonding to take place to lay the foundation for positive future experiences and environmental dispositions. Alternatively, caregivers may want to hold infants close while seeking out new vantage points such as watching and listening to leaves flutter in the trees, hearing chipmunks chirp and watching them scurry, or experiencing the wind howl in a cool autumn breeze. These sensory-rich experiences form implicit memories that will leave a lasting impression on an individual's EID, even though they may or may not be vividly recalled until later in life (Morgan, 2010).

Mistrust in Nature

Once again, a child's ability to form a trusting relationship with nature is highly influenced by family members, sociocultural values and beliefs, and geographical contexts. Unfortunately, not all children have opportunities to build a trusting relationship with nature through shared and supported experiences with caregivers during the first years of life. Thus, educators also play an essential role in helping children establish or reestablish a sense of security and comfort with the natural world.

The following example, told to me by an environmental educator, provides an illustration of how *Mistrust in Nature* may present itself in individual students. The educator took a group of inner city 7-year-old children on a walk in a wetland woods just outside the city limits. Observations of one child's behavior offered critical information for the educator to identify what stage of their EID needed to be nurtured. When the child saw that the trail cut into the dark forest outline, he muttered, "*I am scared,*" physically stopping just outside the wall of trees. In this circumstance, the child clearly needed to revisit the first foundational stage of developing *Trust in Nature*, before he could progress in other stages of EID. After he looked past the wall into the vibrant forest and observed other children comfortably running into the trees, he slowly walked in. Following a short duration of time, he began to loosen up, venturing a little further into the woods with his peers. However, later during the walk, he stepped on a stick, which tipped up behind him and tapped his backside. The child screamed, startled and terrified by the stick, believing that something was on his backside. The educator showed him how the stick had flung up by his own weight. This helped him to calm down momentarily, yet it was evident that because of his lack of exposure and experience in wooded environments, he needed to establish security and a sense of *Trust in Nature* as a foundation for framing future positive encounters with the natural world.

Negotiating Environmental Tensions

EID is fluid, meaning that with each new encounter, new experience, and new context children may need to reestablish a sense of *Trust in Nature* in order to further progress in their EID. While children may establish a sense of trust in one environmental context, a novel situation or experience may present a new dilemma that requires the foundational stage of *Trust in Nature* to be revisited. For instance, while a child may feel comfortable going on a day hike and following a trail, hiking at night or camping overnight in the woods may cause a child to experience a certain degree of discomfort. Specifically, a child may hear a strange noise in the dark night or see a shadow in the trees. Momentarily, a child may experience mistrust, fear, and anxiety. However, this dissonance is essential to EID as it allows children to come to terms with their sense of self in relation to particular environmental features. It is only through establishing a sense of trust with all of nature, both hidden and revealed, that a child will find the strength to overcome anxiety and discover again a sense of comfort and security with the natural world.

Navigating Rosebushes

An example of children's encounters with wild rosebushes in my *Welcome too Our Forist* research project reveals the various ways that children overcame the "outer" dilemma of pokey rosebushes and the "inner" dilemma, or fear and anxiety, that children faced in navigating the prickly vines. From a child's height, the towering rosebushes at or above eye level appeared overwhelming to many of the children. Upon entering the forest, at first sight of the prickly bushes, one child nearly ran into a divider that blocked access to motorized traffic. *"There's prickly bushes everywhere!"* he exclaimed, which he repeated over and over again throughout his exploration. He clearly expressed his bewilderment with the overbearing vines. Another child paced back and forth alone among a cluster of black spruce trees. Talking to himself, he stated, *"I am not scared of the bushes."* Demonstrating a much more intrapersonal disposition, he engaged in an inner debate with himself in order to establish a sense of trust in this new environment. Several teachers asked him if he needed assistance, yet he ignored their requests. Eventually he mustered up the confidence to venture further into the forest, following a group of children nearby. One group of approximately eight children clustered together in a line rather than wandering on their own, moving slowly between and around the bushes. Calling out to one another as they advanced through the forest, more confident peers paused to help their friends when their clothes became caught or snagged in the spikey branches. Others showed their peers who were stalled or delayed by the bushes how to step on top of and over the prickly vines so as not to get tangled or scratched.

In this example, each child approached the dilemma in a different way; their various approaches were informed by their dispositions, social interactions, and previous experiences. While one child felt the need to scream out his anxiety, another quietly talked himself through the situation, and children with more confidence scaffolded their peers in gaining skills to overcome an environmental challenge.

Revisiting Trust in Nature through Environmental Education

During the *Welcome too Our Forist* research project, teachers also helped children who were anxious gain confidence in navigating the rosebushes. The following is part of a video transcript of three children learning with their teacher how to navigate rosebushes during their first day exploring the forest with their preschool class. The teacher, Ms. Tiffany[1], was guiding the children, Rachel (age 3), Tate, and Dominic (both age 4), back to the main trail where the rest of the class was gathering to leave the forest. This excerpt provides an authentic example of how teachers can support and empower children to gain a sense of trust in navigating challenging environmental features.

Dominic moves slowly through the deep rosebushes that are as high as his chest. He reaches out and grabs ahold of Ms. Tiffany's hand. Tate, a short distance away, gets his foot caught between a log on the forest floor and a rosebush.

Tate: *Ouch!*

Tate also reaches for Ms. Tiffany's hand; she assists him in regaining his balance.

Rebecca, who is nearby, walks right into a tall rosebush.

Ms. Tiffany: Don't touch that sticker bush, honey. Yep, that's a big one.
Rebecca: *Ow!*

Ms. Tiffany reaches her hand towards Rebecca who was reaching her hand towards a rose bush.

Ms. Tiffany: If you grab onto that it is going to poke you right through your sweater. See you have it there in your hand. There you go!

Dominic seems to be stalled by a tall thicket of rosebushes, some of which are taller than him. He turns around as if he was contemplating going back. Ms. Tiffany notices Dominic in the bushes.

Ms. Tiffany: Let's go Dominic!

Rebecca and Tate navigate through the bushes more quickly than Dominic who is having a difficult time getting through the rosebushes. Rebecca is singing as she passes through the forest.

Rebecca: *Jingle bells, jingle bells, jingle all the way!*

Dominic's shoe gets caught in a vine and the thorns poke his ankle. He lifts his leg and rubs it.

Dominic: *Ah!*

Dominic hunches over and grabs his leg again, and his back rubs up against a larger rosebush behind him. He cringes his arms together and squeezes between two tall bushes.

Dominic: *I keep getting a prickle. Ah!*

He lifts his leg up and rubs it.

Ms. Tiffany: You need your strong blue jeans.

Dominic: *What's strong blue jeans?*

Dominic is full attention with his arms up high navigating through as best he can. His feet continue to get snagged by the prickly bushes.

Ms. Tiffany: Blue jeans, they are just a little bit tougher pants, or Carhartts. Carhartts would be perfect.

Dominic gets to a clearing and Rebecca is standing there. She giggles.

Dominic: *What?*
Rebecca: *Jingle bells …*

She giggles excitedly and continues through the forest, barely flinching over the rose bushes. Dominic attempts to step over a large bush. He puts his foot on top of it and then gets poked.

Dominic: *Ah!*
Ms Tiffany: Give it a try. Big, big step. Watch.

Ms. Tiffany picks up her foot and shows Dominic.

Ms. Tiffany: If you pick your foot up and step on top of it you can go right through it or you can go around it.

Dominic looks at the bush in front of him and decides to try to get around it. He walks around it, only to encounter another bush.

Ms. Tiffany: Now, look where you are going.

Dominic moves at ease through the forest walking in the spaces between the rosebushes.

Ms. Tiffany: That a boy!

Dominic encounters a few more small rosebushes and cringes as they rub against his leg. Then he stops at a large rosebush that towers above his head, separating him from the main path, his final destination where he will join the rest of his classmates. The rosebush is massive for a child of his size.

Dominic: *Look! I am going to go this way because there is a spike there.*

He points to the rosebush with his finger.

Ms. Tiffany: Look here, Dominic! Look what I found. You might have to go up and over.

Dominic uses his hands to separate two small bushes and walks in between them. In the distance, the other teacher calls to him to join the rest of his classmates.

Ms. Tiffany: He's trying to navigate his way through the forest.

Ms. Tiffany explains to the other teacher. She stands in front of Dominic pointing out the way to go.

Ms. Tiffany: And this is a safe spot. That is a safe spot to hold onto.

Ms. Tiffany points out an old tree branch sticking out of the ground. Dominic reaches out and holds onto it as he climbs over the decaying tree stump and makes his way to the main path where his classmates were waiting for him.

Ms. Tiffany: Look at that, you made it!

In this account of three children navigating the rosebushes we can see how personality, experience, and support from a teacher all play a role in helping children develop a sense of *Trust in Nature*. The rosebushes did not seem to bother Rebecca who moved freely in between the tall vines merrily singing a song. Her behaviors indicated that she felt confident and secure in the forest setting. Tate demonstrated a bit more apprehension and reached out for Ms. Tiffany's assistance when his shoe gets snagged on a rosebush and log. Dominic's movement through the forest is hindered by the prickly vines, which seemed to cause him much fear and anxiety. Ms. Tiffany suggests that Dominic consider wearing stronger jeans for future forest explorations. In this sense, she is looking after his physical comfort by seeing that the forest foliage did not chaff his legs. Additionally, instead of just telling Dominic what to do to overcome the challenges of the rosebushes, Ms. Tiffany first showed him and then challenged him to do it on his own. She modeled how he should lift his legs high and stomp over the pokey branches; she also showed him how to choose the best path. Dominic first practiced with Ms. Tiffany and then he demonstrated his new skills independently. Ms. Tiffany encouraged him with praise for his accomplishments. In this way, Ms. Tiffany scaffolds Dominic in developing a sense of *Trust in Nature*.

Diversity in Trust in Nature

The emergence of *Trust in Nature* occurs within distinct social, cultural, and geographical contexts. Recently, an Alaska Native (Gwich'in) colleague, Pearl shared with me a childhood memory that occurred one summer at her family's fish camp in northern Alaska. In our conversation about language, land, and ecology, she recalled her non-verbal, intuitive exchange with a bear. Pearl described how in her cultural beliefs each individual has a certain association with a particular animal. For her, she felt a connection to bears. One time at fish camp she stayed in a "haunted cabin" at night while all the other members of her family had gone out. She recalled how during the night the dogs started to bark like crazy and the wind started to blow. Pearl went outside to see what the commotion was all about. Stepping outside, she encountered a large bear feasting on the tundra near her cabin. Taking notice of Pearl, the bear paused from his feverous scooping of food into his mouth. For a moment they looked at each other and then the bear continued eating.

"What did you do then? Did you go back inside?" I asked.

"No, I just stood there and watched it eat," Pearl explained.

"Weren't you scared?" I asked.

"I was not scared at all," she explained and after some time had passed the bear went away. Pearl recalled the experience as significant and since that time she has not encountered another bear in the wild. Pearl's strong sense of *Trust in Nature* is deep and pervasive, providing feelings of reassurance and security during that encounter with the bear that to most might yield high levels of fear and anxiety. What I am suggesting is not that one should put oneself in close proximity with a bear, a large moose, or another potentially harmful animal. Rather, I am suggesting that Pearl's deep sense of *Trust in Nature* emerged through a spiritual and intuitive process of listening to the unspoken language of nature and all beings, what could be referred to as being "in tune" with the natural world. This sense is at the very heart of an acute and discerning environmental identity. Pearl listened to something deep inside her sense of self that told of her special relation with bears, provoking questions about how *Trust in Nature* is not only an expression of cultural values and beliefs but also of spiritual conviction.

Indeed Pope Francis (2015) writes about human's special relation with all of Creation. In his encyclical, *Praise Be to You – Laudato Si': On Care for Our Common Home*, he contends that creation is, "like a sister with whom we share our lives and a beautiful mother who opens her arms to embrace us" (p. 9). Drawing inspiration from his namesake, Pope Francis reflects on the communion Saint Francis of Assisi shared with all beings:

> He would call creatures, no matter how small, by the name of "brother" or "sister." Such a conviction cannot be written off as naïve romanticism, for it affects the choices, which determine our behaviors. If we approach nature and the environment without this openness to awe and wonder, if we no longer speak the language of fraternity and beauty in our relationship with the world, our attitude will be that of masters, consumers, ruthless exploiters on their immediate needs. (p. 15)

Similarly, since the beginning of time, Indigenous cultures have acknowledged humans as an intricate part of all of Creation, recognizing the intimate relations that are forged between humans and other natural beings and objects even at the earliest ages. As explained by Kawagley (2001):

> From the outset the newborn is introduced to the voices of the family members, the words of the midwife, the hum of the wind, the sound of the falling rain and the call of the Artic loon. The newborn is already immersed in nature from its first moments of life. (p. 1)

In this way, conceptions of self are formed through immersion in nature where children begin to recognize themselves as part of all that exists. Over time, these interactions become internalized and inscribed into ones' sense of self. This in turn would help to guide how one might interact with and or react to all living beings as exemplified in Pearl's story. While various traditional cultures throughout the world still carry this understanding, Sebba (1991) explained that many families in Western or urbanized cultures tend to alienate children from natural environments, replacing direct sensory-rich nature experiences with encounters with human-made replicas or models. As Sebba (1991) noted, removing children from the natural environment "is accompanied not only by physical but also by psychological separation" (p. 414). As a result of a nature disconnection, children are developing "critical and analytical" dispositions towards the natural environment rather than "adaptive and sympathetic attitude(s)" (Sebba, 1991, p. 414).

In furthering our understanding of EID, or *Trust in Nature* specifically, it is important to consider the cultural, spiritual, and geographical dimensions of how trust develops and evolves. Even more important is working towards bridging understanding between cultures that might be similar to or quite different from our own. Educationally speaking, Ritchie (2014) wrote of a "critical place-based orientation" (p. 49) in which early childhood educators might partner with local Indigenous Elders. She contends that "dialogical interaction with both Indigenous peoples and the local place itself, is seen as a source for interpreting ways of caring deeply for our planet, positioning humans alongside local ecologies as 'co-habitors' of the earth" (Ritchie, 2015, p. 49). In this way, Ritchie (2014) extends an environmental education focus to consider the spiritual well being of children as well as their physical and emotional well being experienced through "sensory, storied entanglement within the inter-relational agency of other animals, plants, insects, and the rest of the [more- than-human] world around us" (p. 50). Indeed, *Trust in Nature* is intra-relation and interrelation, informing how we relate to and how we are related by all living beings.

Future Directions for Research and Application

Can *Trust in Nature* be established later in life? In sharing research about EID with undergraduate and graduate students, classroom teachers and informal educators, several have commented on observing what they perceive to be a sense of *Mistrust in Nature* among teenagers and young people. They have raised the question about how *Trust in Nature* might be established or reestablished with youth or others who may be lacking in their early formative experiences. As described in this chapter, in our contemporary, technologically driven society, children are not always given the opportunity to have nature experiences that support them in developing a sense of trust. In both urban and suburban environments, many children grow up in a concrete world mostly void of wild, ungroomed places. As such, limited exposure to wilder nature may cause children, and even adults for that matter, to experience fear, discomfort, disgust, and anxiety in natural settings (Bixler et al., 1994; Bixler & Floyd, 1997). This fear may prohibit them from venturing just beyond the trail or stepping too far away from the road. Thus, they are not able to positively progress through the subsequent stages of their EID, namely the next stages of *Spatial Autonomy, Environmental Competency,* and *Environmental Action.* Therefore, there is a need for research to examine the process of establishing or reestablishing attributes of *Trust in Nature.* What might this look like? What experiences, specific encounters, and socialization strategies might be useful for establishing trust where it is desolate or lacking? Additionally, in recognizing that *Trust in Nature* is fluid and can and should be revisited continuously throughout life, research might also look at how formative experiences may implicate different levels of trust. In what situations might mistrust prevail? Are there specific encounters when trust is difficult or impossible to establish? What role does education play in this process?

Conclusion

Establishing a sense of *Trust in Nature* is foundational to an individual's EID. Basic trust provides the impetus in which all relationships are formed, including one's relationship with nature. Psychologically speaking, trust provides a sense of security and comfort and a prevailing belief that someone or something is reliable and good. Trust, however, is always situated within a particular social or environmental context and is thus influenced by interactions with other living beings, both human and more-than-human, as well as geographical settings, cultural understandings, and spiritual belief systems. *Mistrust in Nature*, on the other hand, is formed by unfortunate practices, anxious encounters, and a prevailing lack of confidence experienced in nature. Over time and through repeated negative encounters children may grow to feel unsure of themselves while in nature. Educationally speaking, teachers and caregivers play a significant role in supporting children in overcoming environmental challenges and unpredictable and potentially dangerous situations during nature experiences. As such, *Trust in Nature* might need to be revisited, reestablished, or renewed with each new encounter and environmental context.

Note

1. Names of teachers and children are pseudonyms.

References

Bixler, R. D., Carlisle, C. L., Hammitt, W. E., & Floyd, M. F. (1994). Observed fears and discomforts among urban students on field trips to wildland areas. *The Journal of Environmental Education, 26*(1), 24–33.

Bixler, R. D., & Floyd, M. F. (1997). Nature is scary, disgusting, and uncomfortable. *Environment and Behavior, 29*(4), 443–467.

Bowlby, J. (1980). *Attachment and loss: Loss, sadness, and depression* (Vol. 3). New York, NY: Basic Books.

Chawla, L. (1992). Childhood place attachments. In I. Altman & S. Low (Eds.), *Place attachment* (pp. 63–86). New York, NY: Plenum.

Chawla, L. (2008). Participation and the ecology of environmental awareness and action. In A. Reid, B. B. Jensen, J. Nikel, & V. Simovska (Eds.), *Participation and learning* (pp. 98–110). Thousand Oaks, CA: Springer.

Devine-Wright, P., & Clayton, S. (2010). Introduction to the special issue: Place, identity and environment. *Journal of Environmental Psychology, 30*(3), 267–270.

Erikson, E. H. (1950). *Childhood and society* (1st ed.). New York, NY: Norton & Company.

Erikson, E. H. (1972). Eight stages of man. In C. S. Lavatelli & F. Stendler (Eds.), *Readings in child behavior and child development* (pp. 19–30). San Diego, CA: Harcourt Brace Jovanovich.

Erikson, E. H. (1980). *Identity and the life cycle*. New York, NY: Norton & Company.

Francis, P. (2015). *Praise to you laudato si': On care for our common home*. San Francisco, CA: Ignatius Press.

Greenspan, S. I. (1999). *Developmentally based psychotherapy*. Connecticut: International Universities Press.

Groh, A. M., Fearon, R. P., Bakermans-Kranenburg, M. J., Van IJzendoorn, M. H., Steele, R. D., & Roisman, G. I. (2014). The significance of attachment security for children's social competence with peers: A meta-analytic study. *Attachment and Human Development, 16*(2), 103–136.

Hay, R. (1998). Sense of place in developmental context. *Journal of Environmental Psychology, 18*(1), 5–29.

Heft, H., & Chawla, L. (2006). Children as agents in sustainable development: The ecology of competence. In C. Spencer & M. Blades (Eds.), *Children and their environments* (pp.199–216). New York, NY: Cambridge University Press.

Kawagley, A. O. (2001). Parenting and teaching: One and the same. *Sharing Our Pathways, 6*(5), 1–3.

Low, S., & Altman, I. (1992). Place attachment: A conceptual Inquiry. In I. Altman & S. Low (Eds.), *Place attachment* (pp. 1–12). New York, NY: Plenum.

Mistrust. (2018). In *Merriam-Webster Dictionary online*. Retrieved from https://www.merriam-webster.com/dictionary/mistrust

Morgan, P. (2010). Towards a developmental theory of place attachment. *Journal of Environmental Psychology, 30*(1), 11–12.

Phillips, R. G., & LeDoux, J. E. (1992). Differential contribution of amygdala and hippocampus to cued and contextual fear conditioning. *Behavioral Neuroscience, 106*(2), 274–285.

Proshansky, H. M., Fabian, A. K., & Kaminoff, R. (1983). Place identity: Physical-world socialization of the self. *Journal of Environmental Psychology, 3*(1), 57–83.

Ritchie, J. (2014). Learning from the wisdom of elders. In J. Davis & S. Elliott (Eds.), *Research in early childhood education for sustainability: International perspectives and provocations* (pp. 248–265). New York, NY: Routledge.

Scannell, L., & Gifford, R. (2010). Defining place attachment: A tripartite organizing framework. *Journal of Environmental Psychology, 30*(1), 1–10.

Sebba, R. (1991). The landscapes of childhood: The reflection of childhood's environment in adult memories and in children's attitudes. *Environment and Behavior, 23*(4), 395–422.

Trust. (2018). In *Merriam-Webster Dictionary online*. Retrieved from https://www.merriam-webster.com/dictionary/trust?src=search-dict-hed

Wells, N. (2000). At home with nature: Effects of 'greenness' on children's cognitive functioning. *Environment and Behavior, 32*(6), 775–795.

Wilson, R. A. (1995). Nature and young children: A natural connection. *Young Children, 50*(6), 4–11.

Wilson, R. A. (1996). The development of the ecological self. *Early Childhood Education Journal, 24*(2), 121–123.

· 3 ·

SPATIAL AUTONOMY VS. ENVIRONMENTAL SHAME

My Own Special Place

Nobody knows that I am here; I run over the bridge quickly, before the next car approaches, and dive between the bushes and tall grass that conceal the dusty earth path. My path, nobody sees me, I've made it once again and can already hear the water running, carrying my thoughts, my worries away. I continue on past the sandy bank, my own secret beach. Crushed dry leaves beneath my feet, twigs, and trails, at times I lose the tracks, but my path, I know the way; I've come here many times. One last turn where the river bends and the trail is once again uncertain, I push my way between the trees. At last, the bush breaks open and I see the rock, my rock, standing perched swallowed by the running water, smooth and soothing, gently rippling away, my way, to my own special place.

What brought me to that place time and time again? Was it the secrecy, the curiosity, or the need for privacy? It was a refuge from the storms of my childhood life. I am not sure if my mom ever knew about my special place, or my sisters for that matter. It was at least a mile from my house, clear on the other side of the small mid-western U.S. town where I grew up. That's part of what made it special, the fact that nobody could find me, and that no one really knew that it existed. It was mine, all mine, and I was free to be who I was and to explore my feelings. I had so many back then, so many uncertainties and challenges, and I needed a place where I could watch and just listen to the water run.

It was not until many years later that I realized its significance and how my childhood special place contributed to my interest in the natural environment. Now as I watch my own children discover special places I have more understanding. My children have many such places and are constantly constructing and claiming them both indoors and outdoors. They are much younger than I was when I started venturing to my special place. But upon reflection, I had many of them too, but my other places were not necessarily set in a primitive and wild setting. That place held particular importance; it was my most favorite nature spot. The places that my children select are significant too; each one has meaning and purpose in their discovery. Children view places differently than we do as adults, through the lens of imagination and possibility. I never know when they will discover places. We may be on a bike ride when they insist on stopping to explore a pile of tree limbs. I watch and wait amazed as they move, shape, and build the limbs into a new fort. Or we may be on a hike and they excitedly discover a nook carved out beneath a pine tree, or a climbable rock or hill that they feel the necessity to summit. Indeed, carving out, claiming, and discovering special places seems to be a worldwide phenomenon (Kjørholt, 2008; Kylin, 2003; Sobel, 2002).

But what do these special places mean to the lives of the children who find them and how do such places influence children's EID? In this chapter, I explore *Spatial Autonomy* as it relates to children's spaces in the natural world. These are claimed places, independent or shared places that children have discovered and attributed meaning. Such meaning is created, handcrafted, and as unique as the children to which these places belong.

Let's begin by conceptualizing *Spatial Autonomy*. The word "spatial" is defined as "relating to, occupying, or having the character of space" (Merriam-Webster, 2018). "Autonomy" refers to "self-directing freedom" or "the state of self governing" (Merriam-Webster, 2018). So putting these two words together, *Spatial Autonomy*, in its basic form refers to self-directed space, or space where an individual may seek freedom and independence. Free space? If you critically reflect on this meaning, it is possible to imagine how *Spatial Autonomy* is an evolutionary phenomenon shared by all living creatures. For instance, the red squirrels in the far north must claim a territory, referred to as a midden, to survive. Indeed, most mammals acquire a burrow or den of some sort to store food, protect their young, or stay warm. While particular behaviors and preferences differ among species, it is no wonder that humans, as a mammal, are somehow wired to create their own nooks and crannies in nature. We have been doing it for a long time.

Figure 3.1 reveals the basic attributes associated with feeling of *Spatial Autonomy* vs. *Environmental Shame*. The "I statements" listed here are not meant to be exhaustive. Indeed, these inner attributes, or feelings, may be stronger or weaker in various situations. An individual may fluctuate between positive and negative constructs; this is part of the process of negotiating outer and inner tensions inherent in identity formation. *Spatial Autonomy* is strengthened when a child finds a sense of peace and calmness even among the challenges and uncertain elements of nature. Gaining a sense of *Spatial Autonomy* provides children with the freedom to explore their own feelings, thoughts, preferences, and beliefs. Here attributes of *Spatial Autonomy* are maximized when a child recognizes the inherent good of nature. Nature could be perceived as a place of inspiration and wonder. It might also be perceived as a relative and friend. The image in Figure 3.1 shows a child moving away from an adult figure. It is not necessary that this occur independently, in fact, exploration and discovery of place is often shared with another—be it a peer, sibling, or even a parent. What is important is that children have a sense of agency to discover and explore place on their own. In this way, children soon discover that they are a part of everything that exists, an intricate part of nature.

In opposition to *Spatial Autonomy* is *Environmental Shame*. While autonomy brings about freedom and independence, shame may cause children to feel anxious and uncertain of themselves while in nature. When faced with

Figure 3.1. Spatial Autonomy vs. Environmental Shame © Carie Green 2018.

an outer environmental dilemma (e.g., pokey rosebushes), a child may retreat within self or cling to a parent or caregiver instead of moving outward towards discovery. Over time, a child might decide that they would rather avoid nature experiences all together.

In this chapter, we will begin by examining the psychological development of *Spatial Autonomy* through recontextualizing Erikson's second stage of identity development, namely, *Autonomy vs. Shame and Doubt*. We will then continue to build upon understanding of place attachment and identity development theory as it relates to human relations with nature. Like other aspects of EID, emotions play a key role in mitigating one's experience in the natural world. Therefore, a section of this chapter will specifically focus on emotions associated with *Spatial Autonomy* and *Environmental Shame*. Examples from the *Welcome too Our Forist* research project will also be presented to look at the diverse ways in which *Spatial Autonomy* is experienced independently and collectively through childhood play and exploration. The chapter closes with suggestions of future directions for research and application to aid in more deeply understanding the significance of *Spatial Autonomy* in one's EID formation.

Psychological Development and Spatial Autonomy

As children become more mobile, they gain independence through movement and their ability to master new skills. In the autonomy stage of Erikson's theory, toddlers begin to recognize themselves as independent individuals propelled by their ability to satisfy their own needs. With encouragement, toddlers practice and gain a sense of mastery over self-sufficient skills (e.g., feeding themselves, going to the toilet independently). Continued support and encouragement from adults is essential while children practice and acquire new skills. Otherwise, toddlers may become discouraged and doubtful about their own abilities, resulting in feelings of doubt and shame and negative self-worth.

Coinciding with Erikson's (1950, 1972, 1980) concept of autonomy, Proshansky and Fabian (1987) proposed that children gain *Spatial Autonomy* through exercising agency in particular places and the object within these spaces. This manipulation of places and physical objects provides children with a sense of individuality. Similarly, Laufer and Wolfe (1977) argued that privacy, gained through *Spatial Autonomy*, is absolutely essential to a child's psychological development. In addition, they argued that obtaining a sense of privacy is an expression of personal dignity and enhances a child's self-esteem (Laufer & Wolfe, 1977). Experiencing *Spatial Autonomy* in nature has been discussed in the "significant life experience" literature and it has been found that adults who are engaged in environmental professions or actions have noted how such experiences were formative to their environmental dispositions (Chawla, 1999). Research with young children has also shown that children aged three to five-years also expressed an interest in their own places, or micro-spaces, within their home environments (Green, 2013). The places claimed by children for personal or social use were often too small for adults, providing children with opportunities to explore their identity in the natural world through play or sedentary activities (Green 2013). Findings also shed light on understanding middle childhood fort-building behaviors (Sobel, 2002), finding that young children are less likely to construct their own places but rather prefer to use existing natural spaces (e.g., inside bushes or under trees) to gain a sense of autonomy and environmental identity in the natural world. During the toddler years, a child's desire for *Spatial Autonomy* may be demonstrated indoors through climbing inside a box or a laundry basket. In the outdoors, toddlers may crawl or waddle off the edge of a picnic blanket, or explore just beyond the boundaries set by parents.

Human and Place Attachment

In exploring the relationship between place, identity, emotions, and cognition, Morgan (2010) proposed an integrated model of human attachment and place attachment theory. This model considers the "exploration-assertion motivational" processes in which place attachment develops as well as an attachment figure's role in providing comfort and helping children regulate their emotions. Premised on children's preference for natural settings (Hart, 1979) and the role of emotions in developing relationships between people and place (Kaplan & Kaplan, 1989), Morgan (2010) claimed that the outdoor environment arouses excitement in children, causing them to move away from an attachment figure to explore and play. In turn, a child's "place interactions generate positively affected senses of mastery, adventure, freedom as well as sensory pleasure" (p.15). These positive affected senses coincide with the previously described feelings of *Spatial Autonomy*. However, when a child becomes distressed, anxious, or experiences injury in the environment, otherwise referred to as *Environmental Shame*, the attachment motivational system is ignited. The child then seeks the comfort of the attachment figure for emotional regulation and comfort. Once a child's emotional needs are met, they are once again aroused by environmental stimuli, invoking a desire for exploration. Thus this creates a cyclical pattern, resulting "in a 'to and fro' movement between attachment figure and environment" (Morgan, 2010, p. 15).

In order to study the exploration-assertion motivational system, Morgan (2010) interviewed a small group of adults about their childhood place experiences. His findings suggest a common process of place attachment as articulated in the model, namely that place attachment emerges when children gain a sense of *Spatial Autonomy* through recurring positive experiences. Over time, their experiences of places become internalized and become an important part of a their place identity. However, Morgan's (2010) research did not examine the attachment motivational system, that is, a child's return to the attachment figure in times of distress nor did he further explore the idea that some children may draw comfort from places instead of an attachment figure for emotional regulation (Kaplan, 1995).

A House in the Forest

So what role do children's places and adults play in regulating children's emotions in nature? In my own research, I found that depending on the disposition of the child, both caregivers and places, particularly those spaces in nature claimed by children, might serve the purpose of emotional regulation in young children. Below is an example from the *Welcome too Our Forist* research project of how Heather's "*home,*" a black spruce tree, provided her with a sense of comfort and security. Six-year-old Heather filmed the segment while she was wearing a small wearable camera around her forehead, which allowed me to "follow" Heather during her entire play and exploration in the forest. While the entire play sequence was around 25 minutes, I am only including selective parts that highlight the salient aspects of her experience gaining a sense of *Spatial Autonomy*. The scene begins with Heather and another girl, Priscilla preparing "*food*" in their "*house,*" a spruce tree that the girls had found another group of girls playing at in the forest. The first group of girls had called it their house, and Heather had adopted the idea. She led Priscilla in, gathering petals and leaves from the rosebushes near their "*house.*" She explained to Priscilla that they were making "*tea*" and "*jelly,*" which they needed to "survive in our house." Using the tree sap, the girls stuck the leaves on the bark of the tree. They continued their play for about five more minutes until Priscilla got bored and expressed a desire to go explore elsewhere in the forest.

Priscilla: *I want to go.*

Heather: *I am looking for … I found … Here's the bear! Here's the bear!*

Heather attempted to draw Priscilla back into their imaginary play at her "house." It didn't work.

Heather: *Fine. I'll find a different trail and I already found one that looks beautiful.*

Although Heather was reluctant, she went along with her friend. The scene below begins with Priscilla and Heather exploring another part of the forest.

Heather: (suggesting to Priscilla) *If you get hungry in pretend, then I will lead you back to the house.*

Priscilla: *What did you say?*

(She expresses a bit of uncertainty in Heather's comment.)

Heather: *If you are wanting to get back to the house, I can lead you. When you want to go back to the house. I'll lead you.*

Priscilla appears uninterested, but Heather insists.

Heather: *Hmm … I want to go back home. Follow me and we'll be back there.*

Priscilla follows Heather as she walks back to the tree, her pace quickens, and she begins humming. At first, Heather leads Priscilla to the wrong tree.

Heather: *Is it this tree? No, it's not this tree. I think I know where the tree is.*

Priscilla: *I think it's one trail down there.*

Heather: *Me too!*

They eventually find the tree, recognizing the sticky sap on the bark. The video shows how Heather went away from her "*home*" three additional times to explore with Priscilla. Yet each time she eagerly returned to her home. When her teacher whistles and calls for the children to go back to the school, Heather expresses her desire to stay at her house. She hummed to herself as she continued to play with the bark and trees that surrounded the tree. Heather finally tells Priscilla that she wants to stay at her house.

Heather: *I don't want to go. I want to be in the house.*

Priscilla: *Don't you know you're supposed to go? Do you?*

Heather: *Yes, I like it here. And I explored around and realized this could be … well somebody else finded that out and then I finded that out and then I wanted to use this as my house and I explored the farthest place so we'd be alive. For all the food and drinks. So that is exploring …*

Throughout her time in the forest, Heather's desire to stay near her "*house*" grew. She gently hummed to herself each time she returned, outwardly demonstrating a peacefulness that she associated with her special place. In the end, when the teacher called the children to leave, Heather expressed that she did not want to leave her house. She *"liked it here."* She explained that she had *"explored the farthest place so we'd be alive"* and that she did not want to leave.

This observation also shows the diverse ways that children experience places in the forest. While Heather developed an affinity towards the house, Priscilla did not seem to share the same connection. She expressed a desire to explore other parts of the forest and eagerly encouraged Heather to do the same. As well, the other girls who had first discovered the *"house"* had also moved on with their exploration.

Could Heather's house be a place of emotional regulation? There are indications that this may be true. For instance, when she went away from the house she seemed discontent, desiring to return to her house over and over again. Also, Heather had a quiet and reserved disposition. In other words, she was not outspoken, at least among adults, and did not seem to demonstrate any sort of need for a comforting bond with any of her teachers. Perhaps Heather's place connection provides an example of how place can substitute a caregiver or attachment figure within the attachment motivational system (Morgan, 2010).

Discovering a Monster Castle

Let us now turn to another example of *Spatial Autonomy* from the *Welcome too Our Forist* research project. In this example Sergo claimed a *"monster castle"* on a fallen birch tree in the forest. Although two peers and his teacher initially accompanied Sergo to his castle, the place was distinctly his, conceived through his imagination. Unlike Heather, when Sergo experienced distress during his exploration he sought the care of his teacher for comfort. I captured the scene described below because Sergo was wearing a small wearable camera.

> Sergo points to a stand of trees a short distance away and leads his peers and teacher to his *"monster castle"* in the forest.
>
> Sergo: *I like a castle …*
> Teacher: Does this look like a castle?
>
> The teacher points to an area nearby.
>
> Sergo: *It feels like home … Me go … I like home.*
>
> Sergo points to a tree further away.
>
> Teacher: Show me where … show me where it feels like a castle, it feels like home.

Sergo: *Right there.*

Sergo points to the same area of trees.

Teacher: Okay, lead us.

Sergo leads the group down a small trail.

Sergo: *Come on ... I show you where monsters go, there's home.*
Sergo: *Right here!*

[Sergo] He yells to the group.

Teacher: What do you see?
Sergo: *I see a wood, home. Right here is a home.*

Sergo moves towards the fallen tree, climbing up a smaller branch that extends over the trunk.

In this scene, Sergo verbally and non-verbally expressed his excitement and interest in his castle. "*I like a castle. It feels like home.*" Although the teacher suggested an alternative location, Sergo indicated his desire for *Spatial Autonomy* by stating that he was "*making a home.*" Sergo's persistent actions of leading the group towards the "*castle*" revealed his confidence and determination in the environment. As Proshansky and Fabian (1987) explained, gaining a sense of *Spatial Autonomy* supports children in developing their individuality and self-esteem. His teacher supports Sergo in his quest for *Spatial Autonomy* by allowing him to lead the group to his home.

After climbing up the tree, Sergo moves the dead branches up and down, pretending they are "*monsters*" and "*good guys.*" He attributes sounds to the limbs, like "*Rah.*" After some time he decides that he is ready to get down from the tree.

Sergo: *Ahh! Me go there. That's me go there.*

Sergo points to the ground.

Sergo: *Ugh ... I can't see.*

He attempts to get down from the tree, slowly stepping down the high log and holding onto loose branches for stability.

Sergo: *Me go down.*

Sergo looks down. The branch that he is holding is not sturdy; it swings around and Sergo falls off the tree.

Sergo succumbs to his emotions of hurt and *Environmental Shame*; he lays on the forest floor in defeat. With the fall, his poise changes starkly from the confident and assured demeanor that allowed him to successfully climb the tree. He weeps in anguish for several minutes as his teacher makes her way to him. Scared and insecure, Sergo hesitates to move, expressing his emotions loudly through tears and sobbing. The teacher reassures him, "You're okay" and tells him to "stand up." Sergo hesitates, but slowly stands when the teacher reaches his side.

Forming Spatial Autonomy: A Caregiver's Role

Recognizing how to support young children in gaining a sense of *Spatial Autonomy* in the environment is an important aspect of environmental education. In nature, young children become able to experiment with putting distance between themselves and their caregivers or teachers. This sense of physical distance becomes a vehicle for developing self-confidence as children explore the boundaries of independence and separation from adults (Chawla & Rivkin, 2014; Green, 2011). Exposing children to nature is a simple way to support children in gaining a sense of *Spatial Autonomy* in the environment. There are many things to consider in providing children with opportunities to explore their environments, however. While I would argue that wild, ungroomed places provide more rich encounters and opportunities for creativity, outdoor play areas can also be purposively designed to include varied and diverse landscapes with "loose parts" to manipulate and move around in the environment (Mitchell & Evans, 2008). Textured sand, bumpy bark, and soft grass provide rich sensory experiences for children to feel and explore (NAAEE, 2010). Children activate their senses through touching, smelling, and tasting natural objects as they build their perceptions of their environment. Allowing flexibility in movement and encouraging children to pick up objects and explore is essential. Overly shielding children from these experiences may cause them to develop shame and doubt in navigating the natural world.

For example, a small hike may be a tremendous feat for young children, and their caregivers, during the toddler years. However, it can play an important role in strengthening trust bonds in nature and allowing young children to gain a sense of *Spatial Autonomy*. Adults must remain patient and encouraging as children slowly progress on or off trails. For example, Sergo's teacher supported Sergo in his quest for his monster castle. What is important is to recognize children's agency in taking the initiative on their own. Furthermore, caregivers should also allow for environmental experimentation. A child may select a stick and try beating the stick against a variety of objects (e.g., trees, a rock, the soil). This experimentation helps a child begin to see themselves as part of the environment in manipulating and interacting with natural objects and features. Given their physical stature, young children are positioned closer to the ground and perceive the environment very differently than adults (Corsaro, 2015). They may notice tiny insects creeping across the forest floor, small berries, or tiny buds blooming on foliage and vines. Further, an innate curiosity drives children's early exploration and must not be hampered by hurried adults. Similarly, when children become anxious about what they see, adults must help them to recognize the value of all aspects of nature, including those that are more or less pleasing in appearance. Indeed, this growing sense of *Spatial Autonomy* in nature is foundational in building an identity in which children perceive themselves as part of and not separate from the natural world.

Environmental Shame

When children experience *Environmental Shame*, they may feel anxious or uncertain of themselves in nature. Returning to our monster castle example, after the fall, Sergo felt defeated. In fact, he laid flat on the forest floor weeping both in pain and in the disgrace that he felt from the fall, illustrating how *Environmental Shame* might be experienced by a young child when faced with an environmental dilemma. In Sergo's case, he was faced with the dilemma of getting down from the tree, his *"monster castle"* that he had climbed. While confidence and a strong sense of *Spatial Autonomy* gave him the boost to climb up the tree, lack of confidence and skill contributed to his feelings of *Environmental Shame*, which prohibited him from successfully descending down from the tree. If Sergo's fall had been his last experience with nature or if it was part of a recurring pattern of defeat in the face of nature's challenges, then feelings of *Environmental Shame* might be resounding and become the backdrop of his environmental identity. However, shame in the face of unpredictable and challenging environmental dilemmas is natural, and it is how such shame is addressed that is important. Does *Environmental Shame* cause one to fight or take flight, to take root or flee? In the next section we will look at the role that educators play in helping a child overcome *Environmental Shame* and rebuild their sense of *Spatial Autonomy*.

Revisiting Spatial Autonomy through Environmental Education

Environmental dilemmas or tensions are unavoidable. There will always be situations or circumstances when a child, or even an adult for that matter, feels challenged by aspects of nature. Thus, an educator's role is essential in supporting the development of *Spatial Autonomy* and helping children overcome *Environmental Shame*.

Re-climbing His Monster Castle

Let us turn again to the example of Sergo and his monster castle. During his ascent and fall, Sergo's teacher was nearby. She encouraged his quest to find a monster castle. And although there was some risk involved, she permitted him to climb up the fallen tree. However, Sergo had not been taught about the stableness of branches, which contributed to his fall. Sergo had relied on an unstable branch that swung loose and caused him to tumble from the tree. In utter defeat Sego laid on the forest floor, yet his teacher encouraged him to get up and climb the tree again.

Teacher:	Are you ready to go back?
Sergo:	*Yeah.*
Teacher:	Or do you want to climb back on there?
Sergo:	*Yeah … it's scary.*
Teacher:	It's scary. Do you want me to hold your hand this time? Do you want to try again and I'll hold your hand?
Sergo:	*Yeah.*
Teacher:	Okay. Go right back over there. Go right back to the beginning and I will hold your hand this time.
Teacher:	If you want to get up there you can. You don't have to but if you want to, you can and I will hold your hand.

Sergo climbs back on the tree with his teacher holding his hand.

Teacher:	Careful steps, careful steps.

Sergo is whimpering as he slowly climbs up. He reaches the loose swinging branch that he previously held.

Teacher:	Be careful. Those are breakable.
Sergo:	*Why breakable?*
Teacher:	These are breakable, do you remember how easily they break? Sergo, do you want to finish telling me about the dinosaurs?
Sergo:	*No.*
Teacher:	No? Okay. Is this still a castle?
Sergo:	*No.*
Teacher:	No, what is it now?
Sergo:	*I don't know.*
Teacher:	You said it was a dinosaur home.

Sergo:	*No.*
Teacher:	What was it?
Sergo:	*It was me and Charles … and friends.*
Teacher:	You and Charles what?
Sergo:	*Go with friends.*
Teacher:	Oh, I see. But you got back up there when you fell. You can make it a house if you want, a fort if you want, a castle?
Sergo:	*It's me good guys. It's me good guys!*
Teacher:	Okay. Is it still monsters?
Sergo:	*No it's me … good … and guys.*
Teacher:	Oh, it's you and girls and guys.
Sergo:	*No. Good and a guys. It's me and a scary because it's me and it's tree and it's broken and it's me and it's scary and it's dead.*
Teacher:	It's tree and it's broken.
Sergo:	*Yeah and it's broken there and there and there and it's scary.*
Teacher:	Yeah. What happened when it broke?
Sergo:	*Its scary and …*
Teacher:	It's scary?
Sergo:	*Yeah.*
Teacher:	And what did your body do?
Sergo:	*I don't know.*
Teacher:	Well, did you fall?
Sergo:	*Yeah.*
Teacher:	What are you pretending this to be?

She points to the main branch where Sergo is standing.

Sergo:	*I don't know.*
Teacher:	Earlier you said it's a castle. Is it still a castle?
Sergo:	*No.*

With his teacher's guidance, Sergo faced his fears and re-climbed the tree. His teacher held his hand, not forcing but strongly encouraging him to climb back up his castle using "careful steps, careful steps." Sergo whimpered, revealing his anxiousness and insecurity. With support, he returned to the loose branch where he fell and his teacher warned that the branches are "breakable." Sergo asked, *"why breakable?"* and his teacher reminded him that the limbs had broken previously. In this way, his teacher helped support Sergo's emotional regulation by encouraging him to engage in cognitive appraisal of the situation (Hinton, Miyamoto, & Della-Chiesa, 2008). She prompted him to reevaluate the situation that caused him to fall (the breakable branches) and reaffirmed his confidence in his ability to balance up and down the log. Sergo was also encouraged to communicate his fears and learn strategies to overcome his negative experience.

His teacher also invited Sergo to return to his monster castle. However, in his state of *Environmental Shame*, Sergo is no longer interested in the battle of the monsters or dinosaurs and instead he transforms the limbs into good guys. The tree is no longer his castle, a place where he feels confident and brave. It is merely a tree. Yet, in a subsequent book-making activity in the classroom a few days later, Sergo recalled his monster castle with eagerness and excitement rather than with fear and anxiety. This, in turn, suggests that the re-climb, encouraged by his teacher, may have played a role in helping Sergo internally negotiate the negative emotions associated with his experience. If the teacher had not encouraged Sergo to face his fears, would his last impression of his monster castle been that of defeat and despair? Would Sergo have expressed such excitement in recalling his monster castle experience? Perhaps not, since research suggests that humans tend to avoid situations that are associated with negative emotions (Boyer, 2015; Hinton et al., 2008).

Diversity in Spatial Autonomy

Each spring I teach a course in place-based education and we begin the class with a unit on the psychological dimensions of place. I introduce the concepts of place attachment (Manzo & Devine-Wright, 2014), place identity (Proshansky, Fabian, & Kaminoff, 1983), and *Spatial Autonomy* (Hart, 1979; Proshansky & Fabian, 1987) and invite my students to participate in several activities to reflect on their childhood place experiences. In one activity, students map their childhood homes, including furniture, indoor and outdoor environmental features, places in their yard, neighborhood, and beyond. They not only sketch out these remembered places but they also jot down significant memories of each place and the people with whom they shared them. As part of a series of reflections, I ask my students to think about what aspects of their identity might be related to their early place experiences. One student recalled spending much of his time either reading on the sofa or in a small cabin on his family's property, for example. As a young adult, he is still very reflective and reading and quiet moments in nature remain some of his favorites past-times. Additionally, he was pursuing a career as a high school English Language Arts teacher. Thus, there seemed to be a link between his favorite childhood place activity and his career choice.

Another one of my students reflected on a zip line that her father had constructed in her forested backyard. She recalled how, come rain or shine, she spent hours daringly flying down the line and landing in the river that ran along the edge of her property. As she reflected on this place and it's meaning in her life, she attributed these early place experiences to her adventurous and carefree personality. She still challenges herself to conquer new heights whether it be climbing a mountain or carving out new trails with her snow skis. In facilitating this memory activity with students for several years, I have found that no two early place experiences are the same. While there are some similarities, each person's experience of place is unique and influenced by distinct social, cultural, and geographical features. Thus, the way in which childhood place experiences contribute to an individual's sense of *Spatial Autonomy* and EID is also unique.

In her book, *Raising Ourselves*, Gwitch'in writer Velma Wallis (2002) writes about carving out special places in the snow:

> There was no season that we village children did not enjoy. Winter was for digging tunnels and mazes in the snow berms that the village tractor made as it cleared the roads. It was so much fun that we played until we were nearly frozen. (p. 74)

Even in negative forty-degree temperature, Wallis recalls the adventure of creating shared spaces in the snow. The children would go inside to warm up and then be right back out again, not even the darkness of the Arctic North would prevent them from playing:

> At night we played in darkness ... The stars that illuminated the snow kept us company. When the Northern Lights swirled above, we were filled with fear, for we believed that the lights would come down and take us. (Wallis, 2002, p. 74)

In this passage, Wallis revealed a fear of the Northern Lights. Indeed, her writing indicated an inner tension of *Trust in Nature vs. Mistrust in Nature*, when she was faced with an element of nature that she perceived as scary. However, this sense of Mistrust in Nature was somehow overcome as she did not seem to grapple from *Environmental Shame*, that is, withdrawing from play in the outdoor environment.

Future Directions for Research and Application

In the many years that I have studied children's special places, I have come to recognize that no two places are alike. While there may be some shared characteristics, *Spatial Autonomy*, or one's sense of place in nature manifests itself differently in each child. I have also noted how places evolve from the concrete to the abstract, similar to the progression of cognitive development from early and middle childhood to adolescence and into adulthood (Piaget, 1936/ 1952). During the early years, claimed places seem to be much more concrete as children have a need to physically separate themselves from an adult world. Coinciding with Sobel's (2002) findings, in my study of young children's special places, I found that younger children are less likely to modify the physical elements of their places, with modifications rather made through imagination and pretend play (Green, 2013). During middle childhood, children's places are still very concrete; however, children spend a lot more time designing and constructing the physical elements of their place structures (Kjørholt, 2008; Kylin, 2003; Sobel, 2002). In adolescence, the focus shifts altogether to an identity crisis (Erikson, 1980) where children are much more focused on who they are within, rather than separate from, an adult world. Into adulthood, I find that many adults again have the need for special places, which can be physical (e.g., a rock on a cliff) or abstract (e.g., a space in the head experienced through exercise). Having said all of this, there is a need for more longitudinal studies of children's sense of *Spatial Autonomy* vs. *Environmental Shame* throughout an individual's lifespan. In other words, what places hold temporary or situational meaning and what places have longer lasting and enduring impacts? Are there certain types of places that contribute more salient aspects of one's environmental identity than others? How might *Environmental Shame* manifest itself among children with different dispositions and different cultural backgrounds? As mentioned previously, I am interested in further exploring EID in an Indigenous, specifically an Alaskan Native, context. As we critically look at ways in which humans can strive to live more sustainably with our rapidly changing climate and environment, it is important now, more than ever, to understand how children relate with, perceive, and interact with their natural environmental. Therefore, more research is needed to understand children's *Spatial Autonomy*, and how education can be shaped to support that process.

Conclusion

In this chapter, the second progression of children's EID, *Spatial Autonomy* vs. *Environmental Shame*, was presented. While a strong sense of *Trust in Nature* lays the foundation of a child's growing relation to the natural world, *Spatial Autonomy* achieved through discovering a sense of place in nature enables a child to gain a sense of individuality as well as build confidence and self-esteem. Nature provides a space within which children can freely explore their thoughts, preferences, and beliefs. Through such discovery children begin to feel like they are an intricate part of nature. Contrary to *Spatial Autonomy* is *Environmental Shame*, which might cause children to feel tense or anxious in nature. Over time such negative encounters could lead a child to avoid nature experiences altogether. Strong associations of *Environmental Shame* might also cause children to feel unsure of themselves while in nature.

EID is an ongoing and a cyclical process; there is no right or defined way in which an individual should progress. Progression is immanent, however, and can be characterized as negative or positive depending on the circumstances or situations. Although a child may gain a sense of *Spatial Autonomy*, environmental challenges and dilemmas might cause a child to experience *Environmental Shame* as illustrated in Sergo's example. As also illustrated in the examples in this chapter, *Spatial Autonomy* frequently emerges in a social context, yet even so the individuals involved might establish very different orientations and relationships with place. Take for example the case of Heather and Priscilla; while Heather wanted to explore, Priscilla longed to be home near her tree *"house."* Many of the examples I have used here stem from direct observations of children, providing more insight into the life world of a child and the unique role of imagination in contributing to a sense of *Spatial Autonomy*. Adult reflections also reveal diversity in *Spatial Autonomy*. Cultural, social, geographical, and political contexts inform the way in which places are experienced collectively and individually, both for children and adults. As well, caregivers and educators can help to steer children towards positive encounters in the natural world and to overcome environmental challenges. In the next chapter we will look at the third progression of children's EID, *Environmental Competency* vs. *Environmental Disdain*. This progression often goes hand in hand with *Spatial Autonomy* vs. *Environmental Shame*.

References

Autonomy. (2018). In *Merriam-Webster Dictionary online*. Retrieved from https://www.merri-am-webster.com/dictionary/autonomy

Boyer, G. H. (2014). How might emotions affect learning. In S. A. Christianson (Ed.), *The handbook of emotion and memory: Research and theory* (pp. 3–32). New York, NY: Psychology Press.

Chawla, L. (1999). Life paths into effective environmental action. *The Journal of Environmental Education, 31*(1), 15–26.

Chawla, L. & Rivkin, M. (2014). Early childhood education for sustainability in the United States. In J. Davis & S. Elliott (Eds.), *Research in early childhood education for sustainability: International perspectives and provocations* (pp. 248–265). New York, NY: Routledge.

Corsaro, W. A. (2015). *The sociology of childhood* (4th ed.). Thousand Oaks, CA: Sage.

Erikson, E. H. (1950). *Childhood and society* (1st ed.). New York, NY: Norton & Company.

Erikson, E. H. (1972). Eight stages of man. In C. S. Lavatelli & F. Stendler (Eds.), *Readings in child behavior and child development* (pp. 19–30). San Diego, CA: Harcourt Brace Jovanovich.

Erikson, E. H. (1980). *Identity and the life cycle*. New York, NY: Norton & Company.

Green, C. (2011). A place of my own: Exploring preschool children's special places in the home environment. *Children, Youth, and Environments, 21*(2), 118–144.

Green, C. (2013). A sense of autonomy in young children's special places. *International Journal of Early Childhood Environmental Education, 1*(1), 8–33.

Hart, R. (1979). *Children's experience of place*. New York, NY: Irvington.

Hinton, C., Miyamoto, K., & Della-Chiesa, B. (2008). Brain research, learning and emotions: Implication for education research, policy, and practice. *European Journal of Education, 43*(1), 87–103.

Kaplan, S. (1995). The restorative benefits of nature: Toward an integrative framework. *Journal of Environmental Psychology, 15*(3), 169–182.

Kaplan, R., & Kaplan, S. (1989). *The experience of nature: a psychological perspective*. New York, NY: Cambridge University Press.

Kjørholt, A. T. (2008). Children as new citizens: In the best interests of the child?. In A. James & A. L. James (Eds.), *European childhoods: Cultures, politics and childhoods in Europe.* (pp. 14–37). New York, NY: Palgrave Macmillan.

Kylin, M. (2003). Children's dens. *Children Youth and Environments, 13*(1), 30–55.

Laufer, R., & Wolfe, M. (1977). Privacy as a concept and a social issue: A multidimensional developmental theory. *Journal of Social Sciences, 33*(3), 22–41.

Manzo, L. C. (2005). For better or worse: Exploring multiple dimensions of place meaning. *Journal of Environmental Psychology, 25*(1), 67–86.

Manzo, L. C., & Devine-Wright, P. (2014). *Place attachment: Advances in theories, methods, and applications*. New York, NY: Routledge.

Morgan, P. (2010). Towards a developmental theory of place attachment. *Journal of Environmental Psychology, 30*(1), 11–12.

North American Association for Environmental Education (NAAEE). (2010). *Early childhood environmental education programs: Guidelines for excellence*. Washington, DC: North American Association for Environmental Education.

Piaget, J. (1952). *The origins of intelligence in children*. (M. Cook, Tran.). New York, NY: International Universities Press. (Original work published 1936).

Proshansky, H. M., & Fabian, A. K. (1987). The development of place identity in the child. In C. S. Weinstein & T. G. (Eds.), *Spaces for children* (pp. 21–40). New York, NY: Plenum Press.

Proshansky, H. M., Fabian, A. K., & Kaminoff, R. (1983). Place identity: Physical-world socialization of the self. *Journal of Environmental Psychology, 3*(1), 57–83.

Shame. (2018). *Merriam-Webster Dictionary online*. Retrieved from https://www.merriam-webster.com/dictionary/shame

Sobel, D. (2002). *Children's special places: Exploring the role of forts dens, and bush homes in middle childhood*. Detroit, MI: Wayne State University.

Spatial. (2018). In *Merriam-Webster Dictionary online*. Retrieved from https://www.merriam-webster.com/dictionary/spatial

Wallis, V. (2002). *Raising ourselves: A Gwitch'in coming of age story from the Yukon River*. Kenmore, WA: Epicenter Press.

ENVIRONMENTAL COMPETENCY VS. ENVIRONMENTAL DISDAIN

Mozzarella Cheese

"*Is it mozzarella?*" Timothy asks.

"Is it mozzarella? Do you think that's a mozzarella tree?" His teacher responds.

Four-year old Timothy picks up the decaying tree limb, feeling the soft stringy core inside, and exclaims with excitement, "*Yeah! You get cheese from trees, just like paper!*" Children build connections with their environment through past experiences and previous understandings. During the *Welcome too Our Forist* research project, Timothy associated his previous knowledge of mozzarella cheese with the string-like substance that he discovered in the limb of a decaying birch tree. It looked and felt similar, so of course it must be cheese! Such associations are natural as children grow, discover who they are, and learn to read the world around them. Indeed, children perceive their environments with all of their senses, seeing, touching, tasting, smelling, and interpreting the world often quite differently than adults. In the early years, environmental experiences are ripe opportunities for creativity and imagination, where anything is possible. If paper can come from trees, why not cheese? The early conceptions that children make with their environments, although they might be misconceptions, create cornerstones upon which to build future understandings. While the stringy substance inside a tree certainly piqued Timothy's interest, a peer or adult needs to come alongside Timothy in order to expand his ecological understanding.

In this chapter, we examine the third progression of children's EID: *Environmental Competency vs. Environmental Disdain*. Competency is derived from the word "competent" defined as "having requisite or adequate ability or qualities" or "having the capacity to function or develop in a particular way" (Merriam-Webster, 2018). Thus, there is an underlining positive connotation in the word "competency" that one has an ability or capacity. Similarly, some synonyms of "competent" include: "able, capable, good, qualified." The word "competent" is also related to "masterful, experienced, and accomplished" (Merriam, Webster, 2018). Applied to the example above, Timothy is developing the competency, or requisites to understand what could or could not come from trees as well as his capability to deeply explore the forest environment. Hart (1979) defined *Environmental Competency* as the "knowledge, skill, and confidence to use the environment to carry out one's own goals and to enrich one's experience" (p. 225). This definition asserts that knowledge, skill, and confidence all enrich a child's experience of their environment. Certainly, we can easily see how associating the inside stringy substance of a tree limb with mozzarella cheese enriched Timothy's experience of his environment. Indeed, in this chapter, I argue that *Environmental Competency* is exercised through creativity and imagination during the early years plays an important part of EID because of the rich meaningful associations that are established. Who could forget mozzarella cheese? Certainly it made an imprint on me and I am not even the one who discovered it!

Figure 4.1 reveals primary feelings and attributes associated with *Environmental Competency* vs. *Environmental Disdain*. Again, the attributes listed here are not exhaustive, rather they should be seen as a starting point for other environmental identity characteristics that can be measured on a scale from low to high. Additionally, an individual may fluctuate between positive and negative attributes in various circumstances and situations. *Environmental Competency* results in a positive sense of self worth and confidence in one's ability in nature. A sense of well being with nature might inspire an individual's creativity. Such inspiration could take the form of an adventurous game. For instance, in the *Welcome too Our Forist* research project a group of children invented an "*X-marks-the-spot*" game, motivating a non-ending search for stick crossings, which I will return to in a moment. As an individual gets older, creative inspiration might be expressed in art form like illustrations in nature journals or in creating birch baskets.

Creativity and inspiration, however, only fulfill one part of what makes up an individual's *Environmental Competency*. A holistic understanding of nature as provision wherein one recognizes, respects, and acquires skills to live in harmony with the land provides a deep-rooted *Environmental Competency*. This intimate physical, emotional, cognitive, and spiritual relation with nature as subsistence still exists in Indigenous communities all over the world. As a non-indigenous scholar, I struggle for words to explain what has been referred to as a "kinship" relation with nature (Rose, 2005). However, "this alternative view requires sensitivity toward our ecosystems which allows us to be invited by 'nature' into reciprocal interaction within our ecological systems ... Intrinsic to this process is feeling and demonstrating respect for the more-than-human-world" (Ritchie, 2014, p. 51). Nature in this way is recognized as a living entity; someone with this insight develops a strong *Environmental Competency* and an understanding that everything is interconnected, recognizing nature as a part of oneself.

In opposition to *Environmental Competency* is *Environmental Disdain*. "Disdain" is defined as "a feeling of contempt for someone or something regarded as unworthy or inferior" (Merriam-Webster, 2018). While competency brings about security, disdain, applied to an environmental context, may cause a child to feel insecure and disinterested in nature. As disdain takes root, an individual might begin to perceive nature as a place that should be mastered and controlled. In this way, preserving nature is no longer a necessity – the need is replaced with focusing on how nature affordances might be manipulated to serve an individual's needs. Individuals with a strong sense of *Environmental Disdain* would see themselves as separate from nature, failing to respect the reciprocal relationship between human's and other living beings. Over time an individual might place no value in nature, interpreting it as not worthy of consideration with little or no purpose in one's life.

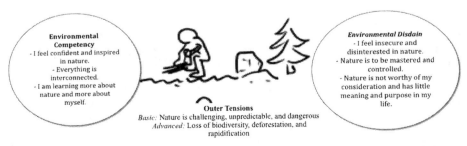

Figure 4.1. Environmental Competency vs. Environmental Disdain © Carie Green 2018.

Outer tensions, or features within an environment that pose challenges and/or are potentially dangerous, play a significant role in shaping and strengthening children's *Environmental Competency*. It is through acquiring skills with which to navigate such tensions that an individual gains a sense of confidence and inspiration in nature. In contrast, when children are generally lacking in skills and knowledge to overcome such tensions, they may develop a sense of insecurity, resulting in disinterest in the environment. The results of advanced tensions, that working on a global level yet have local implications, might include loss of biodiversity, deforestation, and rapidification. "Rapidification," described by Pope Francis (2015) in his encyclical, *On Care for Our Common Home* refers to a "more intensified pace of life and work" (p. 21):

> Although change is part of the working of complex systems, the speed with which human activity has developed contrasts with the naturally slow pace of biological evolution … Change is something desirable, yet it becomes a source of anxiety when it causes harm to the world and the quality of life of much of humanity. (p. 21)

Deforestation and loss of biodiversity monotonize landscapes leading to the extinction of native flora and fauna and the invasion of non-native species. With the loss of primitive landscapes comes the loss of opportunities to connect with and learn about the natural ecology of a place.

Indeed, rapid change and expansion jeopardizes children's development of *Environmental Competency*, particularly the understanding of how everything is interconnected. The global paradigm of rapidification promulgates the idea that nature is to be mastered and controlled, leading to the development of disinterest or disdain towards the environment.

In this chapter, *Environmental Competency* is introduced as an important aspect of a child's EID. First, we will review Erikson's third life-cycle stage, *Initiative vs. Guilt*, in order to specify a child's growing sense of self with the natural world. As with the other stages of EID, examples of *Environmental Competency* vs. *Environmental Disdain* will be presented from the *Welcome too Our Forist* research project as well as other research. In this way, we will explore how children develop *Environmental Competency* alongside peers, teachers, caregivers, and family members. We will also look at the essential role that emotions play in children's encounters with nature, drawing from brain research that shows how positive emotions are linked with competency development and an individual's ability to acquire new skills. A lack of competency, or *Environmental Disdain*, might also emerge in an individual from negative affective responses to outer environmental dilemmas or adverse situations and encounters in nature. The chapter will conclude by examining emerging scholarship on *Environmental Competency* and future directions for research and practice.

Psychological Development and Environmental Competency

In Erikson's third stage of psychosocial development, Initiative vs. Guilt, "a child must now find out what kind of person he [sic] is going to be" (p. 78). With the development of motor skills comes the freedom to explore. Additionally, in this stage children are rapidly acquiring a wider repertoire of language. In this way, children are now equipped to achieve more complex tasks. While in the previous stage of *Autonomy*, a child may have been preoccupied with gaining the skills necessary to accomplish the task of movement, in the *Initiative* stage a child "can forget that he [sic] is doing the walking and instead find out what he can do" (p. 79). Imaginary and dramatic play becomes particularly important during this stage as children explore various social roles and scenarios. Children explore gender roles and the perceived roles of parents and other adults with whom they come in contact. As social skills blossom, children are also more likely to take initiative through play with others. This may lead to "anticipatory rivalry" directed towards other children (peers or siblings) "who were there first" (p. 83) or may occupy a space in which a child's initiative is directed. Erikson (1980) also explains that during this stage, a child's conscience becomes firmly established:

> A child now feels not only ashamed when found out but also afraid of being found out. He [sic] now hears, as it were, God's voice without seeing God. Moreover, he begins automatically to feel guilty even for mere thoughts and for deeds, which nobody has watched. (p. 84)

Thus, Erikson (1980) explains how conscience contributes to feelings of guilt, which can feed a child's inner moral conviction of what is right or wrong. Such convictions might be further enhanced when a child feels like they do not measure up to a parent or caregiver's expectations. Thus, adults play an important role in positively encouraging children's initiative to "make things" (p. 85).

Turning then to an environmental context, initiative, as demonstrated through exploration and play might involve the manipulation of nature affordances for the purpose of symbolically reenacting real world or imaginary contexts. As Erikson's initiative stage is dependent on autonomy, children acquire *Environmental Competency* alongside the achievement of *Spatial Autonomy*. In this way, when children claim or imagine their own places in nature they are able to explore a sense of individuality through their play and the making of their own world. The creative aspects of the natural world positively influence the discovery of self. For instance, in my study of young children's special places (Green, 2011), 4-year-old Nathan concocted "*magic potions*," "*constructed bombs*," and created "*knight weapons*" to thwart off the enemy on the rocks upon the hill behind his apartment. Children might also imagine logs are boats or leaves are cookies during their fantasy play. In this way there is a strong link between *Spatial Autonomy* and *Environmental Competency*.

Environmental Competency is fostered when children have opportunities for repeated and sustained encounters in nature. For example, Barratt, Barratt-Hacking, and Black (2014) described the Redcliffe Children's Centre in the U.K. where "children spend one day a week at one of three forest sites" and emphasis is placed on "exploration, experimentation, observation, problem-solving, prediction, critical thinking, decision-making and discussion experiences" (pp. 234–235). Each new experience may build on previous enactments with other children and the environment and lend itself to creative initiatives. Further, they found that sustained environmental experiences provide opportunities for children to "get to know and develop a relationship with the natural world" (Barratt et al., 2014, p. 228).

Environmental Competency can be garnered through experiences in/from, about, and for the environment (Lucas, 1979). For example, children begin to recognize the creative use of natural objects through spending time in nature, or in places where it's extremely cold, natural objects may be brought indoors for children to creatively manipulate (e.g., willow branches, cedar blocks). The focus in this stage is primarily on learning about the environment. Thus, *Environmental Competency* is developed, as suggested by Hart (1979), through experimentation and manipulation such as what is commonly be seen in "sand or water play" in preschool settings, where children gain understanding of topics such as conservation and the different forms objects take (i.e., solid and liquid). Key, however, to children's identity development is opportunities for children to exercise their own initiative in order to enhance their own experiences (Hart 1979). In finding success in self-initiated tasks, children develop a stronger sense of self and confidence. In contrast, continued failure or discouragement from others when performing such tasks in nature may result in a lower sense of self-worth and associated feelings of *Environmental Disdain*, which could lead to a separation between an individual and the natural world.

X-Marks-the-Spot

Let us now explore an example of *Environmental Competency* in the *Welcome too Our Forist* research project. Here I will introduce the "*X-marks-the-spot*" game invented by the children during their play and exploration in the forest. The children engaged in the game by seeking and finding sticks that crossed in the shape of an X in the forest. Below, is a transcribed scene depicting four boys (Lucas, Grayson, Robert, and Sydney) engaged in the game during their play in the forest:

> Grayson is standing with Lucas who is holding a small stick in his hand near the X-marks-the-spot that they just discovered. Lucas appears to be the leader of the group. He waves his stick at Sydney and Robert who are nearby, inviting them to be a part of the game.

Lucas:	*Hey look – we found tons of X-marks-the-spots!*

> Robert points at two sticks that are crossed upright.

Robert:	*Right there! Right there!*
Lucas:	*Where?*

> Lucas uses his stick to point at the large X that is formed where a heavy birch log leans against a tree.

Lucas:	*Oh yeah, we already spotted that.*
Teacher:	I think ... did you spot that on the very first day, Lucas?
Lucas:	*Yeah.*
Grayson:	*Look it, Lucas. This is X-marks-the-spot! X-marks!*

> Grayson points with his fingers at other sticks on the ground, making an X.

> Lucas uses the stick in his hand to trace the X, confirming Grayson's observation.

Lucas:	*X-marks!*

> The four boys huddle around the new X-marks-the-spot that Grayson found. Robert smiles.

Robert:	*X mark!*
Sydney:	*X ...*
Robert:	*X-cark* [sic]*!*

> Lucas crouches down next to the new X-marks-the-spot and attempts to pick up a large birch log.

Lucas:	*Let's see if I can lift this log up! Ooh, that's heavy!*
Robert:	*Maybe if I could help! I could help.*
Sydney:	*I could help.*

The three boys attempt to lift the log together as a group, eventually getting the log upright and in position.

Grayson:	*You're doing it! Yay!*
Sydney:	*Don't step on that X!*
Robert:	*Here he comes, with the big stick!*

The boys position the stick into an X-marks-the-spot. Robert wiggles the log forward and backward.

| Lucas: | *Robert, stop, I want to keep it like that!* |
| Grayson: | *Can I do something? Can I move it like this?* |

Grayson moves the log forward.

| Grayson: | *Almost, Lucas. X-marks-the-spot!* |
| Lucas: | *Ohhh!* |

Lucas takes a step back and realizes that the two branches/sticks they had been jostling were in the formation of a large, upright X.

| Lucas: | *Hey, X-marks-the-spot!* |

The situation described above represents one of the many "X-marks-the-spot" adventures that the children had in the forest. The game was entirely invented by the children along with its many iterations, including "T-Marks-the-Spot" and "V-Marks-the-Spot." In fact, the game was completely new to the teachers who at first had no idea what the children were doing. As previously mentioned, *Environmental Competency* is enriched through sustained, repeated encounters with nature. Indeed, when Robert shouted "X-marks-the-spot," pointing to two large sticks positioned upright that crossed, Lucas informed him that it had already been spotted and his teacher concurred that Lucas had discovered that particular feature on his first day in the forest. This also reveals how *Spatial Autonomy* and *Environmental Competency* more often than not go hand and hand – Lucas and his friends (and their teacher) had become familiar with and established a connection with a certain area in the forest through their shared game. In the example, we also see how "X-marks-the-spot" evolves from discovered stick crossings that were naturally formed in the forest to the children purposely constructing X's. In this sense, initiative is demonstrated through the invention of a game followed by the manipulation of natural objects for children's own creative purposes (Hart, 1979).

How might we interpret children's EID, particularly *Environmental Competency*, in light of the "X-marks-the-spot" initiative? Lucas's role was distinguished; he appeared to be the leader of "X-marks-the-spot," which grew in popularity among most all of the preschool children. Subsequently, during classroom activities when children were asked to reflect on their forest experiences, Lucas was usually mentioned in other children's reflections and representations of "X-marks-the-spot". Thus, his leading role likely boosted his confidence and self-esteem. As well, Lucas, who appeared to be the "gate-keeper" of this activity, did not accept some of the children's suggestions. For instance, when Robert pointed out the "X-marks-the-spot" that he had found, Lucas shot down his suggestion by stating that he had already found that one. Later when Robert was wiggling the large log in place to form an "X-marks-the-spot," Lucas told him to stop because he wanted it a certain way. While it was not indicated by his behaviors, it is likely that Robert might have experienced self-doubt during this experience, which if frequent and recurring could contribute towards feelings of insecurity and disinterest in nature, or *Environmental Disdain*. However, because Robert's actions were at least somewhat accepted by his peers, his *Environmental Competency* seemed to be fostered, at least in his overall expression of a positive association with the forest environment. All in all, this example shows how frequent and recurring exploration and play in nature can contribute to children learning more about their environment and more about themselves.

Inspiring Environmental Competency: A Caregiver's Role

We have already established that children, from the time they are born, soak in the world around them through sensory-rich experiences. These experiences, good or bad, shape how they feel about and interact with their environment. Equally important to their sensory rich encounters are the attitudes, values, perceptions, and behaviors that are modeled to them by nearby adults. In other words, parents and caregivers play an essential role in shaping children's EID. A few years ago one of my college student's remarked that her child absolutely detested bugs. I asked her how she felt about bugs. "I am scared of them. They are disgusting," she remarked. It was no wonder that her child felt the same way. Through repeated encounters, and likely over time, her mother had modeled that insects were a source of anxiety. This learned attitude and behavior informed her child's *Environmental Competency*, which, arguably, became a source of *Environmental Disdain*. The child devalued bugs, failing to see the reciprocal value of insects and their place as part of the whole. While some insects are widely admired for their beauty and popularized in artistic icons and visuals (e.g., butterflies and ladybugs), and others are well recognized as beneficial (e.g., bees as natural pollinators and honey-makers), most are rated as detestable, pesky nuisances that must be rid of at all costs. Historically speaking, human interventions to rid crops of pesky critters through the application of DDT, for example, had serious implications for wildlife, birds, domestic animals, and humans alike (Carson, 1962). While the widespread application of DDT in the 1950's and 60's was an extreme case of environmental degradation, it was heavily influenced by an anthropocentric mindset that existed then and still persists today. Values passed down from generation to generation, from parents to children, elders to youth, shape how individuals view and interact with nature. We thus must consider what values we pass down as educators and caregivers, what actions we model, and what competencies we teach to children from a very young age. Whether implicitly or explicitly, a caregiver's values, attitudes, and behaviors, unless otherwise convicted, will likely become a part of a child's environmental identity.

Cultivating Strawberries

Let us turn now to an example from *Children's Environmental Identity Development in an Alaska Rural Context.*[1] During Timothy's tour, captured by a wearable camera, Paul's father showed seven-year-olds Paul and Timothy how to cultivate a strawberry plant. The group set out to find a strawberry plant that had been planted on a previous visit to the camp:

Paul's father: Maybe if we are lucky we will get to eat a strawberry.

Timothy: *Yum … strawberries are one of my favorite foods.*

Paul's father: Oh yeah? We have a bunch of strawberries growing in the greenhouse right now.

Timothy: *Uh huh, yeah.*

Paul ran ahead, leading Timothy off the trail to a grassy area.

Paul's father: To your left.

The group shuffled through the tall grass to a small area that had been dug out. A tiny green strawberry plant was growing in the dark clay soil. Paul's father bent down to examine it.

Paul's father: Yeah it looks like it is not doing great—but it is not dead.

Paul's father pointed to a small shoot.

Paul's father: Oh, it sent a runner out and it looks like the runner was unsuccessful in finding a spot.

Using his finger, Paul's father cleared the grass next to the runner. He found another runner.

Paul's father: What about this runner? Nope, this runner was also unsuccessful.
Timothy: *What's a runner?*
Paul's father: Well, the way the strawberries grow, they send out these things called runners, and if it finds good dirt, it will plant another plant.

Timothy watched as Paul's father took the second runner and attempted to push it into the dirt.

Timothy: *Oh, that one too?*
Paul's father: Yeah, this soil is pretty hard.
Timothy: *Should we put some water on it?*

Paul's father did not respond to Timothy's question.

Paul's father: You can see the leaves are starting to turn yellow because it is starting to get cold in the night time. This is going to try to winterize itself soon. And depending on how cold it is in the winter this may or may not live.

Adults model *Environmental Competency* to children through exploration and experimentation as illustrated in the example above. The dialogue revealed that Paul's father grew strawberries in his greenhouse and perhaps he planted the strawberry plant near the camp. Through the demonstration he showed the children how they might redirect a strawberry runner into the soil. Although Paul's father was not certain if the strawberry would survive the harsh conditions of the winter Arctic, he modeled how to cultivate it with care in hopes that it would endure. He discussed how a strawberry plant naturally winterizes itself, suggesting that its survival would depend on the temperature.

Learning to grow food is an important skill for living more sustainably. Gardening is widely recognized as an important aspect of sustainability and environmental education (Williams & Brown, 2012). Thus, teaching children how to grow food in the unique conditions of their local environment serves to strengthen their *Environmental Competency*. Indeed, gardening in the far north is much different than other parts of the world. Likely, Paul is learning firsthand from his father a powerful lesson of how to grow food, both in a greenhouse and through experimenting in the outdoor environment. In this example, we see that Timothy was just as eager to acquire skills for cultivating plants, although he may or may not be learning those skills from his family. In this way, a caregiver's lesson benefits not only his dependent but also his child's peers. Follow-up activities (e.g., revisiting the strawberry plant, comparing a plant grown outdoors with a strawberry grown in the greenhouse) would teach the children *Environmental Competencies* of how to best cultivate strawberries in the Alaskan Arctic. This would serve to strengthen children's initial interests and promote future *Environment Action*.

Through my initial analysis of this situation I noted how modeling cultivation of a strawberry plant might enhance a child's *Environmental Competency*. However, through a second critical analysis, I wondered if there were other implications to consider in introducing a "non-native" species to an environment. I wondered if the strawberry would, perhaps, cause harm to other living beings, flora and fauna of that habitat. While I do not believe that Paul's father or the children meant to cause *Environmental Harm*, I think it is important for us to consider what are the boundaries of our human experimentation in the gaining of *Environmental Competency* with our environment?

Environmental Disdain

So what does *Environmental Disdain* look like and how can it be remedied? Returning to the description of *Environmental Disdain* presented at the beginning of this chapter, one would expect it to emerge as an individual's general disinterest in the natural world. However, as *Environmental Disdain* takes root, an individual might begin to perceive nature as a place that should be mastered and controlled. In this way, nature preservation is not prioritized as necessary or important – the need is replaced with seeing how nature affordances might be manipulated to serve an individual's needs. Additionally, in validating an anthropocentric positioning of nature as a separate entity outside oneself, an individual would fail to fully acknowledge and respect the reciprocal relationship between humans and other living beings.

So what can be done? While a large dose of nature and a significant amount of time is certainly the starting point for addressing what has been referred to as the nature-deficit problem (Louv, 2008), persistent *Environmental Disdain*, characterized as an identity construct, may be more challenging to overcome as it may be deeply inscribed into an individual's psyche. Thus, it is difficult to prescribe an "exact remedy." We can however, consider how fear and stress hinder the recognition and strategic network of the brain, and how such fear conditioning can result in sustained aversive (i.e., avoidance) behaviors. In other words, fear conditioning occurs when an aversive stimulus is paired with a neutral context or stimulus, resulting in an expression of a fear or avoidance response to an originally neutral context (Phillips & LeDoux, 1992). There are many situations in which fear conditioning can develop in nature, particularly through challenging, unpredictable, and even dangerous situations. Let's take for example a child's encounter with a snake (aversive stimulus) when playing in a ditch in their backyard (originally a neutral context). As a result of this terrible fright, the child will no longer go into a ditch or trench when they play outside. This in turn, plays a detrimental role in their environmental exploration, resulting in *Environmental Disdain*, avoidance, and separation from nature.

To address fear conditioning and aversive responses to nature, Hinton, Miyamoto, and Della-Chiesa (2008) suggested a process referred to as cognitive appraisal. Cognitive appraisal is a technique that can be used for regulating negative emotional reactions (Hinton et al., 2008). Through this process, a person would reevaluate the situation as safe and harmless and reaffirm his or her own abilities. Educators play an important role in reducing stress or fear in learning environments through supporting children in communicating their difficulties, establishing an environment where it is okay to make mistakes, and helping children develop strategies to cope with and overcome negative experiences (Hinton et al., 2008). For instance, in the example of a child's fear of ditches, an adult or another more competent individual may encourage a child to verbalize their fears, help them to identify the different types of snakes that inhabit the region, and teach them how to appropriately act when encountering a snake. Equipped with skills and strategies, the child can develop a sense of confidence to overcome fears and anxiety that may have resulted in *Environmental Disdain*.

Revisiting Environmental Competency through Environmental Education

In addition to helping children address their feelings of *Environmental Disdain*, educators should also consider ways in which to nurture children's *Environmental Competency*. Cutter-Mackenzie and Edwards (2013) propose "purposively framed play" (p. 203) as a pedagogical approach for building young children's *Environmental Competency*. Purposefully framed play, as defined by

Cutter-Mackenzie and Edwards (2013), are "experiences in which a teacher provides children with materials suggestive of an environmental/sustainability concept and provides opportunities for open-ended play, followed by modeled play and then teacher-child interaction/ engagement" (p. 203). In this way, beginning with open-ended play supports children in exercising their own initiative. Modeling and discussion serves to expand children's ecological knowledge through connecting their discoveries with existing knowledge about an environmental topic. Open-ended play and modeling, therefore, provide a good balance between offering support and allowing children opportunities to build their own competencies.

Pinky Creek

The following is an example of supporting the development of children's *Environmental Competency* in nature. This example taken from *Children's Environmental Identity Development in an Alaska Rural Context* reveals how two 7-year-old boys discover a "creek" during their open-ended play and exploration at a nature camp.[2] "Pinky Creek" was actually a drainage ditch that diverted water away from the main camp building. However, in exploring the route of the "creek" and considering river flow dynamics the children discovered more about ecological processes and their role in it:

Spencer: *Pinky creek?*

Spencer noted the sign next to a stream of water next to the trail.

Ryan: *Pinky Creek.*

Ryan repeated.

Spencer: *Who put that sign there?*
Ryan: *Hey, this is a creek.*
Spencer: *So, what's this pole come from?*

Spencer noted a white pipe that drained water into the creek.

Ryan: *That's to lead the water from here, this side.*

Ryan pointed out how the pipe extended to the other side of the rock path, near the building.

Spencer: *Oh, it is like a rainage [sic] ditch.*

The two boys follow the "little creek" to a stick poking out of the water.

Ryan: *One little creek, it really starts getting deep here.*

Ryan pointed to a small gully where the water was deep.

Ryan:	*Then it comes through ...*
Spencer:	*... and goes to there.*
Spencer:	*Like a rainage* [sic]*ditch ...*

The boys backtracked, closely examining the course of the water.

| Spencer: | *But there is no water here.* |

Spencer noted a shallow muddy section with little standing water.

| Ryan: | *Yeah, that's because it is low tide through that part. Now, right here you can load a little toy boat. Now, right here it's real deep. Good for heavy toy boats.* |

Ryan bent down and felt the water.

| Spencer: | *Yeah, but here ...* |

Spencer followed the "creek" backwards to where the thick grass made the stream shallow and boggy.

Ryan:	*Its real shallow, good for little ...*
Spencer:	*Only tiny ones.*
Ryan:	*Yeah.*
Spencer:	*If it's big it will drown.*
Ryan:	*It will pretty much hit the bottom.*

In this example, Pinky Creek served as a learning model for the boys to explore human interactions with the environment. Specifically, the children explored the miniature creek to learn firsthand about the different characteristics of a stream and its flow capacity. Through play and exploration of the creek, the children considered what type of boats could be used in shallow and deeper sections of the creek. The model, practically speaking, revealed how humans interact with their environment to divert water away from a building. The lessons learned during their exploration and experimentation may serve to paint the way for future actions such as ecological planning and sustainable development.

Diversity in Environmental Competency

Children's experiences in nature are socially, culturally, and geographically positioned. Thus, when considering the development of children's *Environmental Competency*, it is important to consider how children's understandings and their relations with their environments emerge within various socio-cultural contexts. Indeed, Malone (2016) argued that the literature on connecting/ reconnecting children with nature tends to be informed by a privileged Western, White, middle-class point of view. This viewpoint romanticizes the past by claiming that children used to be closer to nature in previous generations. Along with that, the drive to reconnect children with nature is also based on the assumption that children of contemporary times are nature-deficient (Louv, 2008). While it may be true that the time of many is heavily occupied by technology and structured recreational and/or indoor activities, this deficit discourse paints the picture of a standard view of childhood across the globe that fails to recognize distinct, and perhaps more meaningful, experiences of nature among children in other cultural, social, and geographical contexts (Malone, 2016).

By idealizing one way of connecting with nature, namely through recreation and play, we dangerously glamorize past generations of Western middle-class culture as the ideal when "good parents" allowed children freedom to roam and explore their neighborhoods and environments. Was it not this generation that popularized industrialization, which led, along with human fallacies, to the current state of environmental degradation? Indeed, it is important that we critically examine our assumptions as we consider children's EID. Others have also scrutinized a false dichotomy that positions children as separate from nature. It is risky to assume that through merely "reconnecting" children to nature by way of allowing them more freedom to roam and play, we might begin to resolve our pressing and prevalent environmental issues (Dickinson, 2013; Fletcher, 2016). Certainly, the current state of environmental degradation is much more complicated than that.

So, what other notions of childhood and nature should be considered? First, it is important that we acknowledge that children are not separate from nature. Indeed, how can humans be separate from the Earth since we are mammals who are part of ecological communities whether we recognize that or not? Maybe we will be closer to that in the future should the predictions of humans abandoning the planet made in films like the Walt Disney movie *Wall E* come to fruition, but that hardly seems something we ought to be striving for. Rather than claiming that children are disconnected from a separate entity considered "nature," then, it might be more beneficial to acknowledge that all children are living and breathing natural beings who have and always will be dependent and connected to nature. Furthermore, by examining nature relations through an Indigenous lens rather than an egocentric Western one, we might further examine how children's inter-relational agency with the more-than-human world serves to strengthen their *Environmental Competency* (Ritchie, 2014).

"Good Water to go Fishing"

In another example from *Children's Environmental Identity Development in an Alaska Rural Context*, I found that generally speaking most children living in the village had a strong *Environmental Competency*, both know-how and know-what, about their local ecology.[3] In the extreme climate and geographical isolation of village life much of the community is still reliant on subsistence hunting and gathering food and other resources. Five-year-old Desiree's video tour revealed her subsistence connection. As we ventured down the forest trail, Desiree spotted the high river through the tussock.

Desiree:	*Wow—look at that water!*

Desiree waits until Carter and I catch up and exclaims again.

Desiree:	*Look at this water! This is sea right there. Good water to go fishing!*
Me:	What is that?
Desiree:	*I say good water to go fishing.*
Me:	Good water to go fishing?
Desiree:	*Uh-huh.*

Desiree paused and looked out over the water. Her comments revealed how she associated water with fishing. She called the body of water the "*sea*," which suggested that she had previously fished in the ocean near her village. Desiree led us a little farther down the trail, pointing to the rose hips berries.

Desiree:	*Look what I found.*
Me:	What is it?
Desiree:	*See.*

She pointed to one of the rose hips.

Me:	What is it? Do you know?
Desiree:	*Those are berries.*
Me:	Those are berries?
Desiree:	*Yeah.*
Carter:	*You can't pick them.*
Desiree:	*You can't pick them. They have to get red!*

Desiree indicates her familiarity with another environmental feature. While Desiree pointed out the wild rose hip, she did not pick it, indicating that it was not yet ready. Her friend Carter agreed.

Through this interaction we see a different sort of *Environmental Competency* than previously revealed in the *Welcome too Our Forist* research project. At five-years-old, both Desiree and Carter demonstrated a practical understanding of local food resources whereas Timothy from the *Welcome too Our Forist* research project, who was also five-years-old, expressed a misconception about the source of cheese (i.e., cheese is from trees). Both examples reveal children's inquisitiveness as a guiding factor of their inquiry, but practical experiences in a rural Alaskan context, and likely modeling from competent adults, taught the rural children subsistence knowledge and skills necessary for survival.

On a personal note, living in subarctic Alaska has revealed weaknesses in my own *Environmental Competency*. The intense, and potentially deadly, cold climate of the far north requires one to develop skills for survival unmatched in more temperate climates. During the cold, dark winters, precautionary and proactive measures must be taken into account for every small and large situation. For instance, in the −40 degree temperature, a person could develop frostbite on skin exposed to the cold even for a brief moment. On a more extreme note, absolute reliance on electric power and gas-generated heat could also prove extremely dangerous. Without an alternative source of heat (e.g., wood-burning stove), I have felt vulnerable and scared in my environment during power outages. As well the intensity of heavy snowstorms and high winds have come close to burying me in my quarters for a sustained amount of time. Thus, living in this place of extremities has helped me to recognize the importance of acquiring skills that would ensure my survival should I be cut off from outside resources for certain amounts of time. Beyond developing a craft for shoveling snow, I also need to develop an understanding of the various sources of subsistence and learn the skills necessary for hunting and gathering in my place.

This ancient wisdom of place, passed down from generation to generation in Indigenous and remote communities, has been displaced in most contemporary societies. While there are pockets that have retained a subsistence lifestyle of living off the land, most humans rely on food from the grocer, bananas from Equador, apples from Fiji, and "fresh" oranges from Florida. Amazingly, the science of shipping these products to the far reaches of the world has been somewhat perfected, particularly in the urban centers of Alaska, although there is still some variability in getting such perishable items and other food commodities to Alaska's more rural and remote communities.

Thus, as we consider the development of children's *Environmental Competency* in light of various socio-cultural and geographical settings, it is important to consider the unique conditions in which children's environmental identity is developed and nurtured. Overall, in comparing findings from the *Welcome too Our Forist* research project with findings from *Children's Environmental Identity Development in an Alaska Rural Context*, I found that while children in both places had playful interactions with their environment, a large majority of the children who had grown up in an Alaskan rural village context identified nature as a source of subsistence whereas, children in the *Welcome too Our Forist* research project, most of whom were raised in a more urban Alaskan environment, interacted with nature more as a source of recreation. While both sets of children developed competencies in regards to their environments, their competencies were distinctly informed by family lifestyle and community culture.

Future Directions for Research and Application

So, where do we go from here? In this chapter we have examined the various constructions of children's *Environmental Competency* in different contexts and practices. There is a lot to be learned about what types of competencies (knowledge and skills) that constitute a strong environmental identity. For instance, while I would personally consider myself an environmental advocate and in that sense a person who has a strong disposition to act on behalf of the environment, my recent relocation to the subarctic North has made me recognize the gaps in my ability to fully adapt to the harsh climate of my place. In other words, out of necessity for survival I have had to face my own weaknesses brought about by the challenges and anxiety posed by my environment. While recognition of my weaknesses is the first step, I must now seek opportunities to acquire the skills and abilities that are required for acclimating to my new environment. This has caused me to also critically examine the *Environmental Competencies* that are explicitly or inexplicitly taught to our children. In this chapter, we have noted variations between children reared in a more urban context of the far North, with easy access to modern conveniences, compared to those who are reared in isolated remote villages where subsistence practices still remain an active part of day-to-day life. While children in both situations enjoy nature for its creative benefits (e.g., play and make-believe), the children who depend on the environment as a source of subsistence generally possessed a deeper understanding of their local ecology and how parts of nature fit together. This level of understanding, arguably, is necessary for survival. To this end, future research and practice might include longitudinal investigations of EID in children who have more of a subsistence lifestyle, with a particular interest in the unique experiences and situations that shape and inform these children's environmental identity. Along this line, it would also be interesting to do some comparative studies with children in various cultures to look at how *Environmental Competency* is fostered and nurtured through culture and family. While further inquiry into *Environmental Competency* is certainly a starting point, it is important to consider how competency holistically influences one's willingness to act to preserve or care for the environment. In other words, what competencies are necessary for preparing youth to address the catastrophe of global climate change? Are there certain competencies that must be fostered and what do these look like? These are difficult, perhaps even unyielding questions that may not be easily answered, but a worthwhile endeavor for research.

Conclusion

In this chapter, the third progression of EID, *Environmental Competency vs. Environmental Disdain* was presented. A range of examples reveals that there is not just one way that this aspect of an individual's environmental identity progresses. Rather *Environmental Competency* emerges within various sociocultural and environmental contexts. Additionally, children are active agents in making meaning and developing their own skills in their environments. While children's initial understandings of natural phenomenon may not always be accurate, their meaning making process connects past experiences with new insights. Overarching societal values and family and cultural traditions also influence their understandings. As such, the role of adults and peers is crucial in coming alongside children and helping them learn and develop relevant and meaningful understandings associated with their place. Children's *Environmental Competencies* also appear as creative endeavors; invented games and play activities in nature encourage children to establish their own connections. In this chapter we examined the X-marks-the-spot game as one such endeavor. This lively activity of searching out stick crossings promoted observation where children took note of the environment around them, sought landmarks, and had fun in their adventure. The distinction between *Environmental Competency* and *Environmental Disdain* is not stark; like other progressions of EID an individual may experience one or the other or both at the same time. Certain contexts or situations might momentarily promote feelings of *Environmental Disdain*, but through the process of cognitive appraisal, one can move from the anxiety caused by *Environmental Disdain* to the surety and confidence brought forth by *Environmental Competency*.

In this way, the knowledge, skills, and dispositions gained in relation to the development of *Environmental Competency* sets the stage for an individual to engage in *Environmental Action*. In contrast, *Environmental Disdain* and the disconnection that comes with it might cause an individual to turn away from caring for nature and turn towards the promotion of *Environmental Harm*. While my hope is for the former, we must acknowledge the fickle nature of the human condition. While one might argue for a perfect progression of EID, with no faltering or sways from the positive to the negative, we must understand that these sways or dips from highs to lows, are what strengthen us to rise to the challenges of environmental uncertainty. And perhaps the more tensions an individual encounters the stronger his or her environmental identity will be? Indeed, the stronger we become, the more equipped we will be to face our enduring climate catastrophe.

Notes

1. This example was adapted and reprinted with permission of Springer Nature: Springer, *International Journal of Early Childhood*, Children's environmental identity development in an Alaska Native rural context, Carie Green, 2017. https://link.springer.com/article/10.1007%2Fs13158-017-0204-6
2. This example was adapted and reprinted with permission of Springer Nature: Springer, *International Journal of Early Childhood*, Children's environmental identity development in an Alaska Native rural context, Carie Green, 2017. https://link.springer.com/article/10.1007%2Fs13158-017-0204-6
3. This example was adapted and reprinted with permission of Springer Nature: Springer, *International Journal of Early Childhood*, Children's environmental identity development in an Alaska Native rural context, Carie Green, 2017. https://link.springer.com/article/10.1007%2Fs13158-017-0204-6

References

Barratt, R., Barratt-Hacking, E., & Black, P. (2014). Innovative approaches to early childhood education for sustainability in England. In J. Davis & S. Elliott (Eds.), *Research in early childhood education for sustainability: International perspectives and provocations* (pp. 225–247). New York, NY: Routledge.

Carson, R. (1962). *Silent spring*. New York, NY: Houghton Mifflin Company.

Competency. (2018). In *Merriam-Webster Dictionary online*. Retrieved from https://www.merriam-webster.com/dictionary/competent

Cutter-Mackenzie, A., & Edwards, S. (2013). Toward a model for early childhood environmental education: Foregrounding, developing, and connecting knowledge through play-based learning. *The Journal of Environmental Education, 44*(3), 195–213.

Dickinson, E. (2013). The misdiagnosis: Rethinking "nature-deficit disorder". *Environmental Communication, 7*(3), 315–335.

Disdain. (2018). In *Merriam-Webster Dictionary online*. Retrieved from https://www.merriam-webster.com/dictionary/disdain

Fletcher, R. (2016). Connection with nature is an oxymoron: A political ecology of "nature-deficit disorder" *The Journal of Environmental Education, 48*(4), 226–233.

Francis, P. (2015). *Praise be to you laudato si': On care for our common home*. San Francisco, CA: Ignatius Press.

Green, C. (2011). A place of my own: Exploring preschool children's special places in the home environment. *Children, Youth, and Environments, 21*(2), 118–144.

Hart, R. (1979). *Children's experience of place*. New York, NY: Irvington.

Hinton, C., Miyamoto, K., & Della-Chiesa, B. (2008). Brain research, learning and emotions: Implication for education research, policy, and practice. *European Journal of Education, 43*(1), 87–103.

Louv, R. (2008). *Last child in the woods: Saving our children from nature-deficit disorder*. (expanded version). Chapel Hill, NC: Algonquin.

Lucas, A. M. (1979). *Environment and environmental education: Conceptual issues and curriculum implications*. Melbourne: Australia International Press and Publications.

Malone, K. (2016). Reconsidering children's encounters with nature and place using posthumanism. *Australian Journal of environmental education, 32*(1), 42–56.

Morrison, J. (Producer), & Stanton, A. (Director). (2008). *Wall E* [Motion Picture]. United States: Pixar Animation Studios.

Phillips, R. G., & LeDoux, J. E. (1992). Differential contribution of amygdala and hippocampus to cued and contextual fear conditioning. *Behavioral neuroscience, 106*(2), 274–285.

Ritchie, J. (2014). Learning from the wisdom of elders. In J. Davis & S. Elliott (Eds.), *Research in early childhood education for sustainability: International perspectives and provocations* (pp. 248–265). New York, NY: Routledge.

Rose, D. B. (2005). An indigenous philosophical ecology: Situating the human. *The Australian Journal of Anthropology, 16*(3), 294–305.

Williams, D., & Brown, J. (2013). *Learning gardens and sustainability education: Bringing life to schools and schools to life*. New York, NY: Routledge.

· 5 ·

ENVIRONMENTAL ACTION VS. ENVIRONMENTAL HARM

Litter on the School Grounds

One does not just act for the sake of acting; rather action is a necessary response to a particular context or situation. There was a lot of litter around one school and community in northeastern Arizona where I once taught kindergarten. I took my kids outside once a week to pick up the trash around the school. We brought along several bags for the rubbish and to separate recyclables. While we did our best to clean it all up, it always amazed me how much accumulated over the course of a week. However, in observing the high school youth throw their candy wrappers on the ground while waiting for the bus, I realized this was a deeply embedded behavior, which starkly contrasts with the community's cultural and spiritual values. After school I extended my clean up efforts independently along the side of the road, in fields, by the river, wherever I felt led. The trash really bothered me. I picked up regularly, confident that my efforts could make a difference one section of the roadside at a time.

My class collected aluminum cans to recycle. I remember families used to bring in large garbage bags full of cans and deposit them in the front of my classroom. It did not take long to have a full load to haul to the local recycler, Bill. And, oh my, was Bill a character. His home/business looked like a scene from the old west, rugged wooden buildings among the pink sandy dry desert. He had a conveyer belt that would move the cans up and into a large metal bin. Another middle-aged man sat on the top of the belt to ensure the cans were spread out evenly to properly fill the bin. Bill would buy scrap metal too; while he claimed he was doing his part environmentally, he was certainly making his share of money off of other people's rubbish. Nevertheless, the little bit of money we earned for the class got redistributed for extra supplies and projects.

I opened up my classroom to the community for an earthly cause. I invited a representative from the local Native Environmental Protection Agency to our class to present on waste management issues. We also built a school garden with the help of several community partners. Students planted seeds, watered, cultivated, and maintained their plants. The garden became a daily site for math and science lessons. Students wrote in nature journals, we made tin can stilts, and created mud art. During all these various activities we would have deep conversations about the importance of taking care of the earth, keeping our school clean, and not littering.

After school, I spent a lot of time in the pinion pines, juniper trees, and colorful canyons near my home and school. It is a beautiful place, marvelous and majestic! However, parts of the land were being suffocated under waste, piles and piles of old rubbish and debris. One time I came upon a dump that appeared to be an entire child's bedroom with old toys, carpet, a crib, baby clothes, a dresser … I asked myself, "What happened here? What is happening here? And what should happen here" (Greenwood, 2013, p. 97). I felt a calling to do more.

Pope Francis (2015) calls the Earth "a sister with whom we share our life and a beautiful mother who opens her arms to embrace us" (p. 9) and who "now cries out to us because of the harm we have inflicted on her by our irresponsible use and abuse" (p. 9). But can we hear her cry? And how do we respond? These words held profound meaning to me as I reflected on my own calling, my earthly sojourn. I sat upon a red painted cliff on the edge of the pinion pines looking down into a sharp rock valley that led to an ancient special place, an Anasazi cliff dwelling cut into a granite rock. The valley once teeming with life stood devoid, neglected, and abused. Refrigerators empty and half-opened, bottles of motor oil, old microwaves, a rusty kitchen oven, and other sorts of scrap metal and rubbish scattered along the edge of the cliff. The overlook, with its magnificent view of the high desert landscape, had been defaced. I felt a call to do something, a sense of *Environmental Action* vs.

Environmental Harm. But what was I to do? The answer for me was environmental education.

I remember one of my kindergarten students telling me how she told her family not to use styrofoam because *"my teacher says it's bad for the Earth."* Styrofoam was heavily stocked on the grocer's shelves in the small community where I taught, and it was the cheapest thus most commonly used disposable product for social gatherings. In promoting *Environmental Competency,* I taught my students that styrofoam was a problem because it was not recyclable or compostable, that once consumed and tossed "away," a Styrofoam cup or plate will remain on the Earth for tens of thousands of years. This particular kindergarten student boldly took on *Environmental Action* by verbally conveying her conviction to her family, which may have created a division between what was ecologically sound and her parents' consumer preferences. I am not sure if she was able to change her family's behavior, but taking a stand has to begin somewhere, and why not with the very young?

In the preceding chapters we focused on the foundational attributes (trust, autonomy, and competency) of a child's EID, and here we turn towards considering how these attributes inform a child's will to act. Thus, we examine the fourth progression of one's environmental identity: *Environmental Action* vs. *Environmental Harm.* This culminating stage considers the internal motivators, emotional attributes, and cognitive and moral reasoning that influences one's actions for or against the environment. Action is defined as "an act of will," or "the bringing about of an alteration by force" (Merriam-Webster, 2018). Action can also refer to "the accomplishment of a thing usually over a period of time, in stages, or with the possibility of repetition" (Merriam-Webster, 2018). In the vignette above, my kindergarten class picked up rubbish in the schoolyard on a regular basis, and their actions were progressive, building in momentum and in scale over time. Through repetition the children developed their own will to act. Indeed, two years later when I returned to the community to film an educational documentary about the solid waste (open dumping) issues, the students zealously expressed a desire to go outside and pick up trash on the school grounds again. They had maintained their momentum over time.

Figure 5.1 reveals the basic attributes associated with *Environmental Action* vs. *Environmental Harm.* The listed 'I statements' are not meant to be exhaustive. Additionally, such inner attributes, or feelings, of *Environmental Action* or *Environmental Harm* may be stronger or weaker in various situations. An individual may fluctuate between positive and negative constructs; this is part of the process of negotiating outer and inner tensions inherent in identity formation. The outer tensions presented here are pervasive societal ideals that dissuade care for the natural world, such as consumerism, materialism, and wastefulness. However, it is important to keep in mind that there are a large number of other ideals that may cause an individual to grapple with tensions between *Environmental Action* and *Environmental Harm*. For instance, the ideology of individualism may influence one to harm the environment. One might declare, "I have the right to drive my big truck with a monstrous exhaust pipe because I want to!" which I have found to be a common practice among the rugged cowboy types in the Rocky Mountain college town where I went to graduate school. Never mind the big puffs of exhaust smoke that launched into the air, poisoning the innocent bystander. Drivers of these half-ton steel machines appeared completely oblivious to the environmental and health implications of their actions.

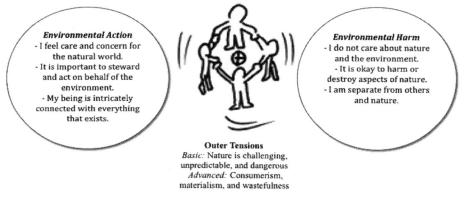

Environmental Action
- I feel care and concern for the natural world.
- It is important to steward and act on behalf of the environment.
- My being is intricately connected with everything that exists.

Environmental Harm
- I do not care about nature and the environment.
- It is okay to harm or destroy aspects of nature.
- I am separate from others and nature.

Outer Tensions
Basic: Nature is challenging, unpredictable, and dangerous
Advanced: Consumerism, materialism, and wastefulness

Figure 5.1. Environmental Action vs. Environmental Harm © Carie Green 2018.

Building on other competencies, a strong sense of *Environmental Action* provides children with the impetus to apply their feelings, thoughts, preferences, and beliefs to action. In other words, strong and pervasive feelings of care and concern for the natural world would subsequently lead an individual to engage in a moral commitment to steward and act on behalf of it. In fulfillment of this progression, individuals would ultimately perceive themselves as intricately connected with everything that exists. The image in Figure 5.1 shows children holding hands around a sphere, illustrating a sense of hope and unity. In this way, *Environmental Action* recognizes that collective efforts are essential to building a sustainable future.

Environmental Harm stands in stark contrast to *Environmental Action*. Harm is defined as "physical or mental damage: injury" or "mischief, hurt" (Merriam-Webster, 2018). Applied to an environmental context, harm would refer to physical damage or injury inflicted on something or someone in nature. While harm can be deliberate and may be associated with an act of will, harm may also occur out of ignorance or complete disregard for the natural world. If one's sense of self is not nurtured towards seeing and acting to preserve the environment, then one is more likely to participate in *Environmental Harm*, either intentionally or unintentionally out of ignorance.

Environmental Harm is like a sword that separates humans from nature, other living beings, and the more-than-human world. In purposively engaging in environmentally harmful behaviors, an individual might declare: "I do not care about nature or the environment." Although children may be equipped with knowledge about environmental issues, the pervasiveness of such issues may cause children to feel hopeless about their ability to take action. Specifically, Zeyer and Kelsey (2013) found that while youth can readily recite the facts regarding the global state of the environment, their response to these concerns portrays a "pessimistic mood," "motive of guilt," and "lack of feeling of control" (p. 207). Similarly, Sobel (1996) described ecophobia, where individuals may feel so overwhelmed by the weight of catastrophic environmental issues that they choose to avoid nature altogether. Under the cloud of ecophobia, a child may wonder what the point is in trying to do something if they feel that nothing they can do will make a difference. Thus, they might endorse throwing rubbish "away" rather than recycling because "it is just one can or one bottle," not being able to understand that every "one" bottle might equate to billions of bottles saved from the landfills each day if every human on the planet took the same measure. While it is quite natural for me to reframe a behavior that causes *Environmental Harm* into an opportunity for action (as the previous statement shows), an individualistic or anthroprocentric mindset seems to pervade contemporary Western culture. More specifically, this individualistic mindset serves to endorse a sense of self that is somehow separated or disconnected from other humans and the more-than-human world.

In this chapter, we will begin by revisiting and re-conceptualizing Erikson's (1980) human life cycle theory. Specifically, we will consider the fourth stage of Erikson's identity development theory, *Industry vs. Inferiority*, and how this can be related to the *Environmental Action* vs. *Environmental Harm* stage of children's EID. We will also examine contemporary research and theory in environmental education and education for sustainability as well as environmental psychology to show how a child's identity emerges as an active agent of sustainable change. As the culminating progression of EID, an array of identity attributes will be considered, including how thoughts, feelings, knowledge, values, moral convictions, behaviors, and responses might be manifested in unique cultural and educational contexts. We will also look at the influence of caregivers as well as the role of formal and informal educators in supporting children's environmental identity formation. I will draw from my own personal teaching experiences, namely, teaching kindergarten and our efforts to promote *Environmental Action* in this particular socio-cultural situation. Vignettes from the *Welcome too Our Forist* research project and *Children's Environmental Identity Development in an Alaska Rural Context* will be integrated throughout as we conceptualize this culminating stage of EID. The chapter will conclude with some thoughts and future directions for considering *Environmental Action* in a diverse, yet ecologically connected world.

Psychosocial Development and Environmental Action

Building on Erikson's (1950,1980) idea of industry, in the later stages of early childhood and into middle childhood, children's attention shifts from self-competencies applied in personal or social situations to those competencies valued by society. This stage of identity development, according to Erikson (1980), is characterized by: "I am what I learn" (p. 87). Children now want to apply their skills and "get busy" doing something (p. 87). All children, Erikson (1980) argued, "receive some systematic instruction" in the development of industry (p. 87). However, not all instruction is of the same nature. Specifically, what is taught and what is learned depends on the context in which such development occurs. In other words, it is dependent on a child's culture, family, teachers, and other social and societal factors.

Learning is also dependent on the will of a child, a child "can watch and try, observe and participate, as their capacities and their initiative grows in tentative spurts" (Erikson, 1980, p. 87). Here "tentative spurts" seems to imply that rather than consistent initiative, there may be other things that distract a child's attention. While a child's will is aroused with an innate desire to learn, which is the foundation of most child development theories, their will to learn and to apply skills may or may not be exercised in every case or in every situation. The word "tentative" suggests that one's sense of industry is accompanied by a focus on what one has set out to do. This focus may be intermittent thus a child's initiative to act must be recognized and supported even in shorter durations, which can be built upon over time because they may or may not yet have the capacity for sustained focus.

Erikson (1980) also argued that children's sense of industry is guided in the middle ground between work and play. This implies that while play can be a driving force of children's activities, the work of *producing things* also begins to steer his or her attention producing things. Erikson (1980) described how children become focused on making things and making things "well" (p. 91). During Erikson's (1980) stage of industry, children also become preoccupied with "*work completion*" in which they pursue with "steady attention and persevering diligence" [emphasis in original] (p. 91).

In opposition to gaining a sense of mastery is the development of inadequacy and inferiority, which could be caused by a number of inner and outer conflicts. For instance, a child might have difficulty adjusting between home life and school expectations, receive limited recognition for their accomplishments, and compare their abilities and skills to peers or family members that are more accomplished. Erikson (1980) asserted that the role of the parent and teacher is crucial to recognizing children's "special efforts" and "special gifts" and providing adequate time and support for the development of children's initiative (p. 92). If feelings of inferiority are not resolved during this pre-adolescent period, due to "lack of inner or outer support of their stick-to-itiveness … lasting harm may ensue for the sense of identity" (p. 93).

Industry and Environmental Action

Extending Erikson's notion of industry, the EID model presumes that children are now ready to apply their previously gained competencies to take action to resolve environmental issues. In other words, *Trust in Nature* provides the foundation of children's environmental identity development, a strong sense of *Spatial Autonomy* and *Environmental Competency* promote children's engagement in *Environmental Action*, which is manifested by their exercise of agency to sustain, conserve, and protect nature. On the other hand, lack of progression in any one of the preceding environmental identity attributes (*Trust in Nature*, *Spatial Autonomy*, or *Environmental Competency*) may lead children to engage in behaviors or develop dispositions of *Environmental Harm* leading to maltreatment of nature (e.g., littering, harming plants or animals). Engagement in *Environmental Action* is an indicator of a strong and healthy environmental identity. As Clayton (2003) explains, "As a motivating force, a strong environmental identity can have a significant impact on guiding personal, social, and political behavior" (p. 46).

So, what does such action look like, particularly in children? Young children's actions are often smaller in scale, but nevertheless still important, and are directed towards fulfilling personal and social goals within their immediate surroundings. For example, in a Swedish preschool study, four and five year-old children demonstrated agency in saving a bird from possible harm (Caiman & Lundegård, 2014). In this particular nature-based preschool, children spend much of their time outdoors thus their familiarity with the environment, resulting from sustained exposure, fostered children's sense of *Trust in Nature*. Further, they have a sense of *Spatial Autonomy* as illustrated in this description of the children in their special place inside the bushes:

> The children creep back into the bushes and sit tightly next to each other. They are quiet for a while. Berit picks up a little chestnut flower, which she places in the hideout, there are some stones and a little grass which mainly consists of both dry and fresh grass. (Caiman & Lundegård, 2014, p. 452)

Also apparent is the children's *Environmental Competency* in using natural objects (i.e., a chestnut flower, stones, and grass) to enhance their experience. As the scene unfolds, the children applied their knowledge about nature to build a bird's nest, which they hoped would soon be filled with eggs. Independent of adults, the children anticipated a potential problem by taking notice of noise coming from a nearby construction site and expressed concern that the place would no longer be good for birds. Next, the children negotiated ways to solve the problem and discussed various solutions before deciding to move the nest to a quieter location. This example provides evidence of children's emerging environmental identities, framed by a collective concern for nature and a strong sense of agency in protecting birds (Caiman & Lundegård, 2014).

While children's small-scale actions may seem small and sometimes even frivolous, it is important to consider the significance of children's initial efforts as these provide stepping stones to bigger and wider-reaching initiatives. In other words, children do not learn how to run before discovering how to coordinate their feet to take their first steps. To be sure, making a difference at a smaller level will have a lasting impact on how a child begins to see himself in relation to the living world.

Discovering a Caterpillar

In another example, taken from my research on *Children's Environmental Identity Development in an Alaska Rural Context*, 5-year-old Sadi discovered a caterpillar on the tundra.[1] In deliberating what to do with the caterpillar, Sadi negotiated the tension between *Environmental Action* and *Environmental Harm*. A crowd of children circled around Sadi, who stood cupping the caterpillar in his hand.

Me:	What is it?
Sadi:	*A caterpillar.*
Me:	Where did you find it?
Sadi:	*Taylor found it.*
Me:	Oh wow! It's a nice one. It's really big.
Sadi:	*I am the one who grabbed it.*
Me:	What are you going to do with it?
Kim:	*Keep it as a pet.*
Sadi:	*Take it to my house and …*
Me:	What do you think it needs?

Sadi did not answer immediately. He held the caterpillar on his finger, examining it closely and stroking its soft body with his other finger. We discussed how caterpillars typically eat leaves and grass and other things living on the tundra. After thinking about it for some time, Sadi decided to set the caterpillar free in its natural setting rather than taking it to his house.

In this example, Sadi wavered between *Environmental Action* and *Environmental Harm*. While he wanted to keep the caterpillar for himself, he considered its natural habitat more fitting. In releasing the caterpillar back into its home, Sadi exercised moral judgment in deciding what was best for another living creature. He learned about the critical role that humans take in harming or preserving others in the more-than-human world, an example of what Ritchie (2014) speaks of as "inter-relational agency" (p. 50).

This *Environmental Action* stage intersects well with all three dimensions of Lucas's (1979) in/from, about, and for the environment model, with a larger focus on facilitating children's agency to act responsibly *for* the environment. Educators and caregivers should continue to take children outdoors and allow opportunities for them to develop personal bonds with nature (Chawla & Rivkin, 2014). This, in turn, can promote a sense of belonging *in* and *with* nature. However, belonging does not necessarily evolve around simplistic peace and harmony; value conflicts and tensions are inevitable in considering how to live sustainably as part of a living whole (Hägglund & Johansson, 2014). Negotiating tensions, that is, inner dilemmas of *Environmental Action* or *Environmental Harm* such as Sadi's dilemma about the caterpillar, shape a child's growing sense of self as an agent of environmental change.

Environmental Harm

Children's EID should not be viewed in stark black-and-white terms. Often there are various shades of gray. These gray shades, more often than not, occur out of ignorance or insufficiently developed *Environmental Competency*. In other words, a child may not know that what they are doing is bad or harmful because they have not learned otherwise. For instance, if Sadi and I had not discussed the situation with the caterpillar, he probably would have taken the caterpillar home, only to find it dead in a jar a few days later. Perhaps Sadi might have given it some green vegetation, he may have even set it by the window for some light, but he might also have neglected to carve air holes in the top of the jar. Thus, the caterpillar, sadly, would suffocate in its new home. While Sadi had no intention of harming the caterpillar, evident in the tender way he handled the creature, harm could have occurred out of a lack of understanding of the caterpillar's real needs. As humans have adapted to a lifestyle where most of their time is spent indoors (Louv, 2008), I would guess that this sort of ignorance occurs frequently due to lack of exposure and misunderstanding of natural processes.

While I am not intending to sound pessimistic, the fact of the matter is that we humans are participating in a relentless race towards environmental destruction with no in end in sight. I wonder how individuals knowingly participate in *Environmental Harm*. Do they somehow justify their immediate need as more important, or do they just not care, or do they feel that their actions will not make an impact (Kollmuss & Agyeman, 2002)? Sadly, this sentiment was recently expressed to me by my own sister who stated, "While climate change may be occurring there is nothing I can do about it anyways" (personal communication, May 2, 2017). To me this cuts to the very core of an identity crisis where humans have attempted to sever their ties with nature. They have lost their external connection as well as an internal conviction that drives one to act to make a difference. We *are* nature. And nature is us! We cannot, or maybe more appropriately, we *should not* turn away from our own. What if all humans on Earth shared this understanding and what if they lived by it? Perhaps, this rearticulates the very purpose of this book, the dire need to consider children's EID with the end goal resulting in a healthy environmental identity that necessitates *Environmental Action*.

"Knock over Trees"

Let us turn, then, from mere speculation to a very real example of *Environmental Harm* from the *Welcome too Our Forist* research project. The situation depicted below was discovered several weeks after it occurred during a review of video footage captured by six-year-old Sonja, with a wearable camera. Much to my dismay when I watched the footage, Sonja had led two other children on an expedition to *"knock over trees"* that involved uprooting and kicking over young birch saplings. Sonja was the originator of the game and also the oldest of the three children involved. It was quite disheartening to watch the incident unfold. I had been in the forest with the children at the time, but I did not realize that the children were engaged in this activity or else I would have discouraged it. The teachers also appeared unaware that the children were desecrating the trees in the forest. Sonja, who was typically very quiet and shy around adults, took on the role of confident leader among the children who participated in her game.

Sonja: *Guys, let's knock this tree over.*

John and Lisa were peeling bark off of a birch tree nearby.

Sonja: *Let's knock it over. Knock it over.*

She starts pushing on the tree. Try to knock it -okay.

John and Lisa joined Sonja in pushing on the tree.

Sonja: *Okay, everybody, everybody start it. Try to do it with all your might!*

All of the children got on the same side of the tree and pushed.

Sonja: *Okay, everybody. I'm pushing super hard!*

The tree creaked.

Sonja: *Oh. Oh, I hear it breaking. I hear it breaking. Everybody.*

The children pushed on the tree. The tree fell.

Sonja: *I did that.*

Sonja mutters to herself a bit surprised that she was able to push over the tree.

Sonja: *Whoa-ho!*

John: *Now we can peel more bark!*

Sonja: *More bark!*

Sonja hops over the fallen tree to the far end and starts peeling bark off its trunk. John joins Sonja, peeling bark near what was the base of the tree.

In this vignette, Sonja was proud of her ability to knock over the tree. She seemed unaware that she had caused *Environmental Harm*. After the tree had fallen the children were excited that they were able to peel its bark. A few days prior, a teacher had taught the kids not to peel bark from trees that were alive and still standing. Applying that knowledge, it seemed that the children reasoned that it was okay to peel the bark from the tree that they had pushed over. In this case, the children's misconceived *Environmental Competency* was applied towards the perpetration of *Environmental Harm*, which often can be enacted out of ignorance or misunderstanding. In this case, a teacher did not witness the children's activities and was therefore unable to explain the ecological implications of knocking over a tree. Additionally, as the scene unfolds we see the children demonstrate an anthroprocentric relationship with nature. In other words, their behaviors endorse a "human-nature binary" expressed through human measures of power and dominion over the more-than-human world (Dickinson, 2013) as illustrated in this excerpt:

Sonja: *Mine.*

Sonja grabs the fallen tree away from others and holds it close.

Sonja: *My tree. My tree. My tree. My tree.*
Lisa: *My tree.*

Lisa takes hold of the other end of fallen tree.

Sonja: *My tree.*
Lisa: *My tree.*

Lisa and Sonja continue to argue back in forth over the fallen tree. John grabs hold of the tree as well.

John: *It's ... all our tree!*

Through their play, the children experience nature as an object to be mastered and controlled. Displaying an anthropocentric mindset, their actions devalued the tree as a living and breathing organism as they argued over ownership. While John suggested shared ownership by stating, *"it's all our tree,"* he nonetheless endorsed human dominance over the tree and participated in the push resulting in its fall, and peeling away the bark that protected it.

Although knocking the tree over and laying claim to it was a game to the children, their assertion over nature is troubling to me. It seemed to represent, in child form, the Western mindset of colonization that claims that land is something to be owned and all living animate and inanimate objects are to be mastered and controlled. And unfortunately, one tree was not enough for the children in this incident as they continued on their quest of dominion and control by searching out a bigger tree:

Sonja: *Let's go look for, let's look for a bigger tree so we can all have it. Like that one.*

She pointed to a tree in the distance that was larger than the one that fell.

Sonja: *Okay, that would be super hard to knock down. Let's look for one that is little that we can knock down.*
Lisa: *Yeah.*
John: *But, but for all of us ...*
Sonja: *I see a tiny, tiny one.*

Sonja ran past John and Lisa who were pushing on another leaning tree.

Sonja: *Oh, this one's super easy to knock over.*

John and Lisa stopped pushing on the leaning tree and walked over near Sonja.

Sonja: *This is so easy to knock over.*

Sonja pushed on the stump.

Sonja: *This is so easy ... well kind of ... like that.*

She pushed the stump over.

Sonja: *That was easy.*

The work of knocking over trees appeared to get easier with each one that the children overturned. Unfortunately, the game did not end with the second tree. There were many more that were uprooted that day in the forest. While the body of literature from the "children in nature" movement advocates for unrestricted free play in nature (Louv, 2008), this incident makes clear that we, as environmental educators, parents, caregivers, and the like, need to critically consider *what* types of nature experiences we are encouraging and endorsing. While I am uncertain if the *"knocking over trees"* game occurred during children's subsequent visits to the forest, I imagine that the excitement and feelings of accomplishment that accompanied such tasks made it attractive to do again. What we do not want in our zealous quest to get kids outdoors is to endorse behaviors that, if not corrected and redirected, will lead to a hardening of the human-nature binary that endorses dominion and control.

Revisiting Environmental Action through Environmental Education

Considering the above example of *Environmental Harm*, we can see why guided education *for* the environment is so important. It is not enough just to send kids outdoors for free play and exploration, although this certainly has its benefits (Russell, 1999). Adult involvement in children's play activities, to some extent, could help prevent reoccurring episodes of *Environmental Harm*. Perhaps not so many trees would have been uprooted and discarded had an adult redirected children during the *"knocking over trees"* game. However, intervention must be done with sensitivity, not in a way that is judgmental and could lead to *Environmental Shame*, but in a way that encourages children to think critically about their interactions. For example, during the first few forest excursions in the *Welcome too Our Forist* research project, the children and adults harvested chaga, a mushroom-like fungus, that grows on birch trees in northern boreal forests. In recent years, the fungus has become popular for its medicinal value. Unfortunately, what I observed with the collection of chaga in this project quickly got out of hand. After a teacher assistant over-filled her pockets with the mushrooms, she encouraged the children to do the same. The children, with the encouragement of that particular teacher, broke chunks of chaga off the trees wherever they could find it, carrying it out of the forest. I became concerned that stripping all the trees of chaga, and teaching the children to do it as well, was unsustainable and potentially caus-ing *Environmental Harm*. I knew the matter needed to be addressed, and with sensitivity. Thus, on a later visit to the forest, I gathered the children under the trees and told them the story of the Petrified Forest in Arizona and how over the years this National Park has lost much of its splendor. Over time, the hundred of thousand of annual visitors to the park have picked up a piece of petrified wood and put it in their pockets, and over time the ancient trees that lay upon the high desert plateau have become fewer. Applying this story to the forest situation, we discussed that in taking material from the forest we take away what makes it beautiful and special to others. The children listened attentively and during subsequent visits to the forest the children and teach-ers quit picking the chaga off the trees. A few weeks later, a child retold the story in the classroom, relating it to his own activities. This, perhaps, was an indication that he had internalized the story and it had affected his perception and behaviors.

This example shows how reflection plays an important role in helping children discover what it means to act for or against the environment. Birdsall (2010) suggests that educators should teach three important components of action: (1) planning or learning how to act; (2) taking action; and (3) reflecting on action. In a sense, thinking about action and taking action occurs continuously, both implicitly or explicitly, in environmental settings. However, after action reflection is something that needs to be explicitly guided. While children may say to themselves, "I am going to pick the chaga off the tree," and then put chaga in their pockets, they may not ask the questions, "Should I pick the chaga? What am I going to do with it? And would picking the chaga provide a conceivable benefit that outweighs harm?"

Action has long been at the center of environmental education. Beginning with the 1977 Tbilisi Declaration, environmental education includes an emphasis on problem-solving and pro-environmental behaviors (UNESCO, 1978). International and national educational guidelines and standards emphasizing environmental care and sustainability have informed efforts around the globe: The *United Nations Decade of Education for Sustainable Development (2004–2014): International Implementation Scheme* (UNESCO, 2005); Singapore's Ministry of the Environment and Water Resources' *National Green Plan* (MEWR, 2012); the New Zealand Ministry of Education's (1996) *Te Whāriki: Early Childhood Curriculum*; the Australian Department of the Environment, Water, Heritage, and the Arts' *Australian Sustainable Schools Initiative* (DEEWR, 2008); and the North American Association for Environmental Education's *Early Childhood Environmental Education Programs: Guidelines for Excellence* (NAAEE, 2010) to name a few. These guidelines are shaped around encouraging children "to investigate, analyze, and respond to environmental changes, situations, and concerns" (NAAEE, 2010, p. 37).

Conceptualized this way, *Environmental Action* might occur within the private or public spheres (Short, 2010). Private behaviors stem from personal awareness of values and are demonstrated through consumer or ecosystem actions (e.g., buying "green," picking up litter, recycling) while public behaviors tend to result in collective engagement and are frequently demonstrated through political or activist outcomes (e.g., participating in an environmental campaign). Both types of behaviors reflect *Environmental Action*. The former represents ongoing personal efforts and the latter is geared toward larger societal resolutions.

Chawla and Cushing (2007) proposed that children's sense of agency to act responsibly for the environment should move beyond the private sphere (e.g., recycling and turning off lights at home) to the public sphere of community problem-solving (Chawla & Cushing 2007). Both private and public actions are important aspects of being an empowered change agent; as Chawla and Cushing (2007) explained, "children and youth need to take personal ownership of the issues that they work on, choosing personally significant goals and integrating action for the common good into their sense of identity" (p. 448). Relatedly, in referring to environmentally responsible behaviors, Heimlich and Ardoin (2008) argue that educators should avoid teaching specific behaviors but rather emphasis should be placed on critical thinking, which, in turn, can empower and motivate action.

Additionally, Phillips (2011) proposed action as a measure of active citizenship where educational experiences provide children with opportunities to express their opinions and make decisions. Likewise, Mackey (2014) described collaborative efforts for creating "cultures of sustainability" (p. 181) in which children, educators, family, and community members come together to address relevant issues within their communities. However, it is important to recognize that the scale of such action, particularly with young children, should be personally, socially, and culturally relevant, so situated within schools, neighborhoods, and other familiar settings. Indeed, a holistic approach where young children engage in agency and are supported to bring about change has been noted by several early childhood education for sustainability scholars (Caiman & Lundegård, 2014; Davis, 2010; Davis & Elliott, 2014; Ji & Stuhmcke, 2014; Phillips, 2014).

Environmental Action centered on solving existing environmental problems might be considered a *reactive* approach to environmental education. While arguably teaching children how to seek solutions to environmental issues is important given the catastrophic state of our ecological system, it is also necessary to consider *Environmental Action* as an internal attribute of a child's identity, a way of being, interacting, and responding to the world. *Environmental Action* as an identity attribute entails a strong sense of belonging where one recognizes oneself as a part of nature and nature as a part of oneself. Many scholars have discussed this notion of "belonging" through various cultural lenses. Belonging, as posited in the literature, provides the legitimacy for children to act with others in constructing a sustainable future (Davis & Elliott, 2014). The Maori of Aotearoa/New Zealand use the term *mano whenua* to talk about belonging, and Mackey (2014) described this approach as a form of critical place-based educational that supports children in gaining an appreciation of and relating to the various cultures and histories that make up their place. Without specifically using the term "belonging," Ritchie (2014) discussed the importance of inter-relational agency that encourages children to see their actions in relation to "other animals, plants, insects, and the rest of the [more- than-human] world" (p. 50). Similarly, Alaskan Yu'pik scholar Angayuqaq Oscar Kawagley (2010) described human belonging in nature as a metaphysic where nature is recognized as a living and breathing extension of oneself:

> Not only are humans endowed with consciousness, but so are all things of the environment. The Yupiaq people live in an aware world. Whenever they go they are amongst spirits of their ancestors, as well as those of the animals, plants, hills, winds, lakes, and rivers ... Because nature is a metaphysic, Yupiaq people are concerned with maintaining harmony in their own environment. (pp. 73–74)

In maintaining harmony with nature, the Yupiaq people are taught "never to do harm to, abuse, or even make fun of animals," plants, or other living beings (Kawagley, 2010, p. 75). Such misuse and disrespect of a certain plant or animal would cause the spirits of that being to not renew its kind again. Through this lens, this explains why certain species have become endangered or become extinct (Kawagley, 1999). In this way, *Environmental Action* is framed around spiritual reciprocity and human relations with all living beings. This proactive, rather than reactive, approach to *Environmental Action* discourages wastefulness and taking more than what is needed. It also demands conscious care and stewardship of nature.

Diversity in Environmental Action

In considering *Environmental Action* as an internal attribute and an extension of oneself, let us examine a situation from *Children's Environmental Identity Development in an Alaska Rural Context*. This interaction was captured by nine-year-old Tabor through a wearable camera. While picking wild blueberries in the low bush tussocks of the Arctic tundra, Tabor muttered to himself, "*Berry combs aren't good for tundra!*" He leaned close, examining and picking the delicate blueberries on the short, stubby bushes of the hillside overlooking the sea. Through his inner dialogue, Tabor revealed his applied understanding and skills (*Environmental Competency*) as a conviction for sustainable action. In his past experiences picking "*blues*," Tabor learned appropriate measures for picking certain types of berries on certain types of terrains. Tabor's declaration revealed his awareness of his surroundings and his desire not to harm his environment. Tabor is from a small rural Alaskan Native village of approximately 700 people. Located off the coast of the Bering Sea, the village is off the road system, over 500 miles away from the nearest city. Out of necessity, as well as in accordance with cultural and family traditions, many of the children in the village engage in a subsistence lifestyle from a very young age. These children acquire a rich ecological understanding of their environment through firsthand experiences. Outdoors in their environment, many can readily identify the various types of berries in the bushes, in the forests, and among the tundra. They distinguish and point out currants, rosehips, high bush cranberries, low bush cranberries, blueberries, and blackberries, among other plants. They know when they are ripe, how to pick them, and their favorite way to prepare and eat them. And in the case previously described, Tabor expressed how to conserve the tundra that produced his favorite kind of blueberry. Growing up in an environment in which families readily depend on the flora and fauna for subsistence means these children established a deep relationship with the more-than-human world. This sense of belonging, as Davis and Elliott (2014) argue, provides legitimacy for children to act in sustainable ways. Tabor's inner dialogue shows how he had internalized a way of being in relation to his natural environment.

Future Directions for Research and Application

So where do we go from here? Research on *Environmental Action* has primarily been informed by a reactive approach to resolving environmental issues. However, as Sobel (1996) proclaimed over two decades ago, children are being burdened with the weight of the world and the heaviness of catastrophic environmental issues much too early. What children need, Sobel (1996) argued, is to first establish a connection with their natural world. This proclamation, as well as research in the significant life experience literature (e.g., Chawla, 1999), has played a key role in prompting the "children in nature" movement and provided a foundation for much North American environmental education research. However, as argued in this chapter, simply sending kids out to play and muck around in nature is not enough. Without proper guidance, children may resort to orientations that actually promulgate *Environmental Harm*. In more recent years a growing emphasis on Early Childhood Education for Sustainability (ECEfS) has emerged around the globe (see Davis & Elliott, 2014). This research emphasizes the role of children as cultural change agents in problem-solving and critically enacting a new tomorrow. This work has also begun to emphasize Indigenous ways of knowing and dialogue with local Indigenous elders to promote a deeper understanding of place (Mackey, 2014; Ritchie, 2014). However, more research is needed to understand the internal identity constructs of *Environmental Action*. Questions that guide inquiry on affective responses, moral reasoning, ways of being, and ways of knowing that orient children towards a more sustainable view of the world would provide us a better understanding of children's EID in diverse contexts.

Conclusion

In this chapter *Environmental Action* was presented as an identity construct internalizing and informing environmental behavior. While environmental problem-solving has long been presented as an end goal of environmental education, and a healthy outcome of EID, we have seen through the examples presented how misconceived *Environmental Competency* can perpetuate *Environmental Harm*. Thus, *Environmental Action* should be more than a reactive approach to addressing prominent environmental issues. Rather *Environmental Action* should be recognized as an identity construct where behaviors are informed by a deep sense of belonging and recognition of self in relation to everything that exists. It is not enough to simply send children out to explore and play in nature, although this certainly has its benefits. The danger is in the perception of nature as merely a site for recreation, without critical reflection on how even play in nature might befit *Environmental Harm*. Moreover, an over-romanticized notion that life was so much better a few decades back when children were free to explore nature at their disposal paints a false hope of a simple solution to the human-nature misconnection. Indeed, those "glory" days were also subject to environmental neglect and abuse, evident in the whistle blowing of Rachel Carson's (1962) *Silent Spring*. Overarching societal tensions of consumerism, materialism, and wastefulness have been around for a long time and have and continue to impact the way in which children view nature as a disposable commodity. Additionally, the looming troubles of global environmental concerns, such as climate change, has caused a pervasive sense of hopelessness among adults who feel that there is nothing that can be done (Kelsey, 2016). Thus, it is even more important to consciously turn our attention, our *sustained* attention, towards holistic EID, in modeling ways of being that help children to see themselves as part of everything that is alive and that every small step, each small token of care for nature matters.

Note

1. This example was adapted and reprinted with permission of Springer Nature: Springer, *International Journal of Early Childhood*, Children's environmental identity development in an Alaska Native rural context, Carie Green, 2017. https://link.springer.com/article/10.1007%2Fs13158-017-0204-6

References

Action. (2018). In *Merriam-Webster Dictionary online*. Retrieved from https://www.merriam-webster.com/dictionary/action

Birdsall, S. (2010). Empowering students to act: Learning about, through and from the nature of action. *Australian Journal of Environmental Education, 26*, 65–84.

Caiman, C., & Lundegård, I. (2014). Pre-school children's agency in learning for sustainable development. *Environmental Education Research, 20*(4), 437–459.

Carson, R. (1962). *Silent spring*. New York, NY: Houghton Mifflin Company.

Chawla, L. (1999). Life paths into effective environmental action. *The Journal of Environmental Education, 31*(1), 15–26.

Chawla, L., & Cushing, F. D. (2007). Education for strategic environmental behavior. *Environmental Education Research, 13*(4), 437–452.

Chawla, L. & Rivkin, M. (2014). Early childhood education for sustainability in the United States. In J. Davis & S. Elliott (Eds.), *Research in early childhood education for sustainability: International perspectives and provocations* (pp. 248–265). New York, NY: Routledge.

Clayton, S. (2003). Environmental identity: A conceptual and an operational definition. In S. Clayton & S. Opotow (Eds.), *Identity and the natural environment* (pp. 45–65). Cambridge, MA: MIT Press.

Australian Children's Education and Care Quality Authority (ACECQA). (2013). Guideline to the National Quality Standard. Online. Available: http://files.scecqa.gov.au/files/National-Quality-Framework-Resources-Kit/NQF93-Guide-to-NQS-130902.pdf

Department of the Environment, Water, Heritage, and the Arts (DEEWR). (2008). *Australian Sustainable Schools Initiative (AuSSI): A Partnerships Statement for the Australian Government and the States and Territories*. Canberra: Commonwealth of Australia.

Davis, J. (2010). *Young children and the environment: Early education for sustainability*. Port Melbourne, Victoria, Australia: Cambridge University Press.

Davis, J. & Elliott, S. (2014). *Research in early childhood education for sustainability: International perspectives and provocations*. New York, NY: Routledge.

Dickinson, E. (2013). The misdiagnosis: Rethinking "nature-deficit disorder". *Environmental Communication, 7*(3), 315–335.

Erikson, E. H. (1950). *Childhood and society* (1st ed.). New York, NY: Norton & Company.

Erikson, E. H. (1980). *Identity and the life cycle*. New York, NY: Norton & Company.

Francis, P. (2015). *Praise to you laudato si': On care for our common home*. San Francisco, CA: Ignatius Press.

Greenwood, D. (2013). A critical theory of place-conscious education. In R. B. Stevenson, M. Brody, J. Dillon, & A. E. J. Wals (Eds.), *International handbook of research on environmental education* (pp. 93–100). New York, NY: Routledge.

Hägglund, S., & Johansson, E. M. (2014). Belonging, value conflicts, and children's rights in learning for sustainability in early childhood. In J. Davis & S. Elliott (Eds.), *Research in early childhood education for sustainability: International perspectives and provocations* (pp. 38–48). New York, NY: Routledge.

Harm. (2018). In *Merriam-Webster Dictionary online*. Retrieved from https://www.merriam-webster.com/dictionary/harm

Heimlich, J., & Ardoin, N. (2008). Understanding behavior to understand behavior change: A literature review. *Environmental Education Research, 14*(3), 215–237.

Ji, O., & Stuhmcke, S. (2014). The project approach in early childhood education for sustainability: Exemplars from Korea and Australia. In J. Davis & S. Elliott (Eds.), *Research in early childhood education for sustainability: International perspectives and provocations* (pp. 158–179). New York, NY: Routledge.

Kawagley, A. O. (2010). Alaska Native education: History and adaption in the new millennium. In R. Barnhardt & A. O. Kawagley (Eds.), *Alaska Native education: Views from within*. Fairbanks, AK: Alaska Native Knowledge Network, Center for Cross-Cultural Studies, University of Alaska Fairbanks.

Kelsey, E. (2016). Propagating collective hope in the midst of environmental doom and gloom. *Canadian Journal of Environmental Education, 21*, 23–40.

Kollmuss, A., & Agyeman, J. (2002). Mind the gap: Why do people act environmentally and what are the barriers to pro-environmental behavior? *Environmental Education Research, 8*(3), 239–260.

Louv, R. (2008). *Last child in the woods: Saving our children from nature-deficit disorder.* (expanded version). Chapel Hill, NC: Algonquin.

Lucas, A. M. (1979). *Environment and environmental education: Conceptual issues and curriculum implications.* Melbourne: Australia International Press and Publications.

Mackey, G. (2014). Valuing agency in young children: Teachers rising to the challenge of sustainability in the Aotearoa New Zealand early childhood context. In J. Davis & S. Elliott (Eds.), *Research in early childhood education for sustainability: International perspectives and provocations* (pp. 180–193). New York, NY: Routledge.

Ministry of the Environment and Water Resources (MEWR). (2012). *The Singapore Green Plan 2012: Beyond Clean and Green Towards Environmental Sustainability* (Online). Singapore: The Singapore Government. Online Available http://app.mewr.gov.sg/data/ImgCont/1342/sp2-12.pdf.

North American Association for Environmental Education (NAAEE). (2010). *Early childhood environmental education programs: Guidelines for excellence.* Washington, DC: North American Association for Environmental Education.

New Zealand Ministry of Education. (1996). *Te Whāriki. He whāriki mātauranga mō ngā mokopuna o Aotearoa: Early Childhood Curriculum.* Wellington: Learning Media.

Phillips, L. G. (2011). Possibilities and quandaries for young children's active citizenship. *Early Education and Development, 22*(5), 778–794.

Phillips, L. G. (2014). I want to do real things: Explorations of children's active community participation. In J. Davis & S. Elliott (Eds.), *Research in early childhood education for sustainability: International perspectives and provocations* (pp. 194–207). New York, NY: Routledge.

Ritchie, J. (2014). Learning from the wisdom of elders. In J. Davis & S. Elliott (Eds.), *Research in early childhood education for sustainability: International perspectives and provocations* (pp. 248–265). New York, NY: Routledge.

Russell, C. L. (1999). Problematizing nature experience in environmental education: The interrelationship of experience and story. *Journal of Experiential Education, 22(3)*, 123–128.

Short, P. C. (2010). Responsible environmental action: Its role and status in environmental education and environmental quality. *The Journal of Environmental Education, 41(1)*, 34–54.

Sobel, D. (1996). *Beyond ecophobia: Reclaiming the heart in nature education.* Great Barrington, MA: Orion.

United Nations Educational, Scientific and Cultural Organization (UNESCO). (1978). Intergovernmental Conference on Environmental Education: Tbilisi (USSR), October 14–26, 1977. Final Report. Paris: UNESCO.

Zeyer, A. & Kelsey, E. (2013). Environmental education in a cultural context. In R. B. Stevenson, M. Brody, J. Dillon, & A. E. J. Wals (Eds.), *International handbook of research on environmental education* (pp. 206–212). New York, NY: Routledge.

· 6 ·

METHODOLOGIES AND METHODS FOR ENVIRONMENTAL IDENTITY DEVELOPMENT RESEARCH

Research is ...

See where you are going.
Play.
Listen.
Looking at things.
Read, write, see what you are doing,
Look at dead moths, or butterflies, or dead moose.
A calendar, write the days.

The steps of research are like ...

Finding direction.
Cross the road.
Stairs. Walk upstairs, walk downstairs.
Ask someone what you want to know about.
Eat supper.

Children say the darndest things! During our first classroom meetings in the *Welcome too Our Forist* research project, I invited children to define what it means to do research. The statements above illustrate some of their ideas.

While their statements are certainly cute, they are also very insightful, and for the most part accurate. Indeed, a researcher would likely *"eat supper"* countless times while engaged in the research process. Additionally, why couldn't research be so much fun that it feels like play? It certainly can and should be when it involves young children. And, yes, there is a lot of reading and writing involved in research, sometimes feeling like it is too much some days. And if a researcher is well organized they will indeed likely keep a calendar to write down what they might do on each day. As well, *"looking at things"* and *"see where you are going"* are important steps to take if you want to discover anything new.

In the *Welcome too Our Forist* research project, our research team (consisting of myself and three undergraduate research assistants) set out to explore what it means to engage young children as active researchers in all aspects of the research process, including: 1) choosing a research topic and posing questions; 2) selecting data collection methods; 3) collecting data; 4) analyzing and interpreting data; and 5) presenting findings. Children, as illustrated above, have a broader understanding of what steps research might involve. Nevertheless, it is important to begin with understanding what children know about research in order to build understanding *with* children, as opposed to *on* children.

The methodology and methods presented in this chapter are grounded in child participation.[1] But beyond just including children's voices and perspectives, the strategies presented here were designed purposively to engage children as the researchers in studying their own environmental experiences. In this way, children's agency is promoted in order to discover the way in which their environmental identity develops and forms. Figure 6.1 provides a conceptualization of my approach in studying EID. Basically, the idea is that engaging children as researchers promotes agency. Promoting children's agency leads to a revelation of their ideas, attitudes, values, beliefs, feelings, actions, and preferences *in, about,* and *for* their environments. These attributes represent their identity constructs, which provide insight into their EID. Identity development, however, is highly influenced by children's past understandings and experiences, social, cultural, and familial contexts, individual personalities and dispositions, and geographical/ environmental settings. Thus, just as children's EID is influenced by various factors, the ways in which we might engage children as active researchers is also dependent on the research context. Thus, while this chapter aims at outlining various approaches that can be used, such approaches should be recognized as fluid and dynamic. Qualitative in nature, inductive child-initiated methods must be malleable to the expressed desires and needs of young children. Prescribed frameworks are not appropriate in considering that children perceive and act upon the world much differently than adults.

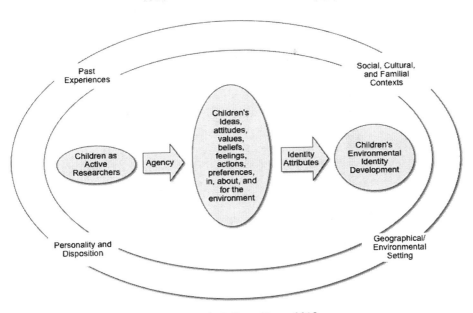

Figure 6.1. Concept map for EID research © Carie Green 2018.

In this chapter, we will begin by outlining the various orientations towards research involving children. We will turn to the literature to understand what is meant by research *on*, *with*, and *by* children approaches in order to make the case for more child-centered methodologies. Then, I will describe some of the innovative methods and strategies used in the *Welcome too Our Forist* research project, including the advantages, challenges, and opportunities of each approach. Lastly, as in other chapters, I will suggest some future directions and practices for exploring young children's EID.

Research Approaches Involving Children

In recent years, researchers in childhood studies have turned towards participatory frameworks aimed at involving children in some aspect of the research (James, 2009). And in the field of environment education, this is no exception.

Indeed, environmental education researchers have begun to recognize the importance of involving children and youth in the research process (Barratt Hacking, Cutter-Mackenzie, & Barratt, 2013; Green, 2015). Children have the right to have their voices heard; their views and perspectives matter. The United Nations Convention on the Rights of the Child (UNCRC) set precedence for children's citizenship and rights of participation recognized under international law (United Nations, 1989; 2005). However, assumptions promulgated under the UNCRC are not without contention. Scholars have interrogated the interplay between globalized discourses on children's participatory rights and how national and local notions of participation are interpreted in various social, cultural, political, and economical contexts (Kjørholt, 2007). Others have noted tokenistic forms of participation where children's participation is manipulated or enacted at a superficial level (Hart, 1997). Still others have argued that the possibilities and limitations of participation should be informed by what children have to say about their participation, recognizing the importance of dialogue with children (Graham & Fitzgerald, 2010). Children from around the world have also voiced what their rights should entail, including rights of: 1) participation and liberty to "express opinions, be listened to, make choices"; (2) protection and "to be treated fairly"; and (3) provision of care and a healthy environment (Taylor & Smith, 2009, p. 171).

Research *on, with*, and *by* Children

In the *International Handbook of Research in Environmental Education*, Barratt Hacking et al. (2013) made a case for involving children as researchers of their own environmental education experiences. They described a continuum of methodological approaches for research involving children from traditional research *on* children to alternative research *with* or *by* children. This continuum can be used to interpret the ontologies, epistemologies, and methodologies that inform environmental education childhood research. Specifically, research *on* children approaches derive from positivist and objective paradigms, and in many cases experimental designs or traditional measures employed in fields such as developmental psychology (James, 2009). Such studies view children as objects of research, perceiving children as "human becomings" not yet developed into fully competent and functioning adults (Lee, 2001, p. 7). Adults are positioned as the primary interpreters of children's experiences and behaviors, with little to no input from children. In perceiving children as incapable of understanding, researchers often elect not to disclose to children that a study is underway. Furthermore, researchers embracing this type of methodology often consider children vulnerable and in need of protection from complex problems or issues (Duhn, 2012). In other words, they are likely to avoid the risks and uncertainty associated with critical frameworks and action-oriented methods of research.

As researchers move along the continuum towards research *with* children approaches, children are more likely to be listened to and acknowledged. Informed by contemporary sociological understandings, children are considered "people worthy of study" whose actions can contribute and make a difference in society (James, 2009, p. 34). Research falling under this approach is more likely to be informed by the United Nations Convention on the Rights of the Child (UNCRC) (United Nations 1989, 2005). Article 12 under the UNCRC established respect for the views of the child, asserting that children have the rights to "participate in all matters of relevance to them," including research (Barratt Hacking et al., 2013, p. 438). Child-friendly data collection methods are commonly used to encourage active participation, such as artwork, photography, and other interactive activities. However, research *with* children approaches are still primarily "adult-led" with findings represented by means of "adult interpretations and understandings" (Barratt Hackling et al., 2013, p. 439; see also Green, 2015). Similarly, Hart (1997) warns against tokenistic forms of participation where children's participation may be manipulated or enacted at a superficial level. In other words, although children's voices are recognized as significant and important, and efforts have been made to include their perspectives in the research, as of yet environmental education research involving children has not yet fully engaged children in directing the goals and course of the project, including posing the research questions and analyzing and interpreting findings (Green, 2015).

In contrast, research *by* children approaches aim to engage children as the researchers or co-researchers of a project. In this approach, children are viewed as competent social actors, "active in the construction of their own lives and the lives of those around them and of the societies in which they live" (James & Prout, 1990, p. 8). Ideally, children would lead all steps of the research process including posing a research question, determining data collection methods, collecting and analyzing data, and interpreting and presenting findings. Rather than leading the inquiry, the researcher takes on a supportive role, stepping away from their own notions of how the study should be done in order to allow children's questions and ideas to emerge naturally. With young children, such approaches have been rare. While many efforts have been made to engage children as active participants in the data collection process, none have yet to explore what it means to engage children as co-researchers in posing a research question, analyzing data, and interpreting findings (Green, 2015).

Additionally, while some have engaged children in project-based problem-solving, particularly around sustainability issues (Ji & Stuhmcke, 2014; Mackey, 2012; Phillips, 2014), these problem-based frameworks emphasize a co-constructive process among children, teachers, families, and community members in addressing local problems. While these "holistic approaches" promote children's engagement in critical inquiry, democratic practices, and community involvement, often the line is blurred between child and adult perspectives. In other words, findings are presented as one perspective with little or no distinction between the perspectives of children and the perspectives of adults. In my own research involving young children, I found that children often perceive the world very differently from adults (Green, 2013), thus this blurring of perspectives is problematic. In recognizing children as active agents of culture and change, environmental education researchers must critically reflect on whose views and agendas are enacted and further problematize whose voices are favored (adults'?) and whose are marginalized (children's?) in "collective" orientations. One way to address this is to implement a variety, or mosaic, of data collection methods in order to provide multiple means for children to share their perspectives (Clark, 2005). Thus, this chapter supports the notion of researchers combining a variety of child-driven methods in the study of children's EID.

Furthermore, ethical questions have also been raised regarding research involving young children, including ensuring child assent, maintaining confidentiality and protection, and the power imbalances between the adult researcher and child participants (Conroy & Harcourt, 2009; Einarsdóttir, 2007; Punch, 2002). While the issue of control has long been debated among qualitative researchers (Lincoln, Lynham, & Guba, 2011), this issue is particularly contentious in research involving young children because of their subordinate positioning in an adult world (Corsaro, 2015). Indeed, the power imbalance between the adult researcher and child participants can greatly impact *what* data is collected and *how* it is interpreted. Some have proposed relational approaches to address ethical concerns, emphasizing the importance of establishing trust with children, creating a comfortable environment, and providing children with choices in participation (Clark, 2005; Einarsdóttir, Dockett, & Perry, 2009; Green, 2012; Parkinson, 2001). However, relational approaches still rely on the direct involvement of adult researchers in the process and, depending on the relationship established, the researcher's involvement will inevitably influ-ence the way children act and behave. In this chapter, then, I want to share an approach I have been using called Sensory Tours (Green, 2016), which I have found helpful as a method to remove the researcher from being directly involved in the data collection process so that I can better explore children's authentic interactions with peers and their environment.

Others have suggested the use of video as a means to "pick up different sorts of voices" along "with a range of images" (Haw, 2008, p. 192). However, control issues have also been raised regarding the use of video data with young children. For example, Robson (2011) discussed the tension of remaining unobtrusive when filming preschool-aged children through the use of tradi-tional video equipment like a handheld camcorder. Additionally, traditional video data is still framed through the researcher's point of view since they are choosing what to film; as Robson (2011) reminds us, "images are never neutral; they are literally and socially constructed" (p. 186). For me, then, I wanted to find ways to involve children in the process of collecting and inter-preting video to increase their participation and I have found Sensory Tours paired with Video-Stimulated Recall Discussions to be helpful in increasing child-directed participation in image-making and interpretation.

Most importantly, adult researchers must be reflexive and adaptable, that is willing to change and adjust things according to children's desires and needs. Each child is different, as well as each encounter and situation; what works for one may not work for another. Therefore, intuitive strategies should be implemented in such a way as to recognize the diversity of children, envi-ronments, and social contexts (Green, 2012). Doing so is not always easy, but it is necessary if we are truly seeking to honor children's participatory rights and perspectives.

Participatory Methods for Researching Children's EID

In this chapter, we will examine several methods for exploring children's EID. Six methods will be discussed: Bookmaking, Sensory Tours, Video-Stimulated Group Discussions, Art-Making, Role-Playing, and Building a Model. While some methods might be used for data collection alone, others can be used to engage children in multiple parts of the research process, including brainstorm-ing initial topics, choosing methods, data collection, analysis and interpreta-tion of findings, and presentation. These methods are informed by participatory and phenomenological methodologies. Phenomenological research is aimed at the nature of experiences and is derived from a "careful description of ordinary conscious experience[s] of everyday life" (Schwandt, 2015, p. 234). Phenomenological meanings are derived from "perception (hearing, seeing, etc.), believing, remembering, deciding, feeling, judging, evaluating, and all experiences of bodily action" (Schwandt, 2015, p. 234). Rejecting scientific realism, reality is interpreted as subjective and is recognized as socially constructed between observers and actors and speakers and listeners (Schutz, 1967). In this way, children's natural world socialization, that is, how the environment shapes young children and how children shape their environment, can be interpreted through their bodily actions and expressive discourse.

Bookmaking

Bookmaking can be used in various forms throughout a research project in order to document what took place and to create a visual that children can revisit time and time again. In the *Welcome too Our Forist* research project, we used three forms of bookmaking. First, we used a *Research Big Book* to document children's engagement in the project from beginning to end. Second, we used a book titled, *Our Data Collection Methods* that was purposely created to invite children to choose their data collection methods. Third, all the children were invited to create their own book page to show and tell about their own experiences in the forest. The children's book pages as well as pages from the *Research Big Book* were compiled into a final book to showcase children's engagement in the project. Throughout the project, bookmaking was used for data collection, analysis, interpretation, and presentation.

Research Big Book. Bookmaking was introduced to children at the beginning of our research to collectively capture their ideas and to document what they did together. In this way, the project becomes their story: *for* them and *about* them. It is a means to engage the children in the making of the research. We used a large flipchart to create a *Research Big Book* of their research. To begin, I said, "The first thing that we are going to talk about is research. How many of you have heard of research? What is research?" By starting with what children know about research, children's ideas were built upon from the inception of the project. Additionally, a *Research Big Book* can be used to explore each step of the research process and the topics that children select to explore. Each time children work on the project they can add new pages to the book, including photographs, quotes from their discussions, and drawings that they create to represent their experiences. In our experience of the *Welcome too*

Our Forest research project, children expressed excitement in revisiting pages that they had previously created. These pages were used to extend and build upon their previous understanding.

Our Data Collection Methods. A book can also be used to engage children in selecting their own data collection methods. We created a book titled, *Our Data Collection Methods* that provided children with examples of child-friendly, age-appropriate data collection activities (see Figure 6.2). The activities (methods) included were informed by a review of research approaches in early childhood environmental education (see Green, 2015) as well as novel approaches scarcely discussed in the literature. Photographs of the children participating in the various activities were included in our book to make it more meaningful to the children involved in the project. The book was presented to groups of children and each child was given an opportunity to look through the methods and choose the ones that most interested them. After the children selected their preferred data collection methods, we designed centers in the forest to facilitate the activities. Four data collection centers were developed, Drawing and Painting, Building a Model, Role-Playing, and Sensory Tours (GoPro), each of which will be discussed in greater detail in subsequent sections of this chapter.

Figure 6.2. Pages of the book *Our Data Collection Methods*.
Source: Author.

Bookmaking for Data Analysis and Interpretation. Bookmaking can also be used as a method to engage children in data analysis and interpretation of their environmental experiences. In our *Welcome too Our Forist* project we worked one on one with each child, inviting children to make a page to share what they remembered and liked best about the forest. Children were presented with an 11 X 17 sheet of paper, photographs, and a variety of craft supplies (markers, stickers, glue sticks, stickers, and glitter paper). The bookmaking activity can be introduced to children by stating: "Today you are going to make a book page about your experience in the environment. I have some picture of you and artwork that you made. You can decorate your book page however you wish."

As children create their book page, researchers can invite children to describe what they were doing in a photograph or what they remembered about an experience. Children can also be invited to draw pictures to illustrate their experiences. Below are some probes that can be used to engage children in data analysis and interpretation:

(1) Tell me about your experience in the [name of environment]. What do you remember? What did you like best? What was your favorite thing to do?
(2) I am going to show you a photograph. What do you find interesting in this photo? What were you doing?
(3) Tell me about what you notice (qualitative description).
(4) Is there anything that you can count? (quantitative description)
(5) Would you like to draw a picture of your experience in the [name of environment]?
(6) Is there anything else that you want to share about your experience?

Bookmaking as a data analysis activity provides an opportunity for children to reflect on and validate their interests in selected topics. For instance, in the *Welcome too Our Forist* project, Peter drew and talked about bugs during his bookmaking activity:

Researcher:	Do you remember what you did when you were in the forest?
Peter:	*Yes.*
Researcher:	What did you do?
Peter:	*I was trying to look for beetles.*
Researcher:	Beetles ... oh wow!
Researcher:	Do you want to draw anything about your experience in the forest?

Peter: *Uh-huh.*

Peter takes the lid off a marker and begins to draw.

Peter: *How about I draw a bug. This one is a bug. It has two legs.*
Researcher: I like that bug.
Peter: *And that one's a ladybug.*

He draws a second bug on the other side of his paper.

Researcher: That's a ladybug?
Peter: *Uh-huh.*
Researcher: And what did you do with a ladybug?
Peter: *I don't know. I went and found one.*
Researcher: Did you find lots of ladybugs in the forest?
Peter: *Yes.*

During the bookmaking process, Peter had the opportunity to interpret data, in this case a photograph, that was created from a Sensory Tour collected during a previous stage of the research. He interpreted the photograph of himself looking for bugs in the forest. He also drew a picture of bugs, demonstrating that he had internalized the search for bugs as an important part of his forest experience. Applying the dimensions of EID, Peter had clearly moved beyond the *Trust in Nature* stage in that it seemed that he felt secure enough to explore nature and gain a sense of *Spatial Autonomy*. The bookmaking activity also demonstrated Peter's growing sense of *Environmental Competency* as he revealed his developing understanding about bugs.

Although Sensory Tours and the use of wearable cameras will be further described in the section that follows, here I will briefly touch on how wearable cameras can be used during the bookmaking process. Specifically, children were invited to wear a small camera (i.e., a GoPro) on their forehead while constructing their book page. The camera revealed children's interests or what drew their attention. For example, the camera showed when Peter looked towards the door, or when he became distracted by a peer or another object in the classroom. The wearable camera also revealed Peter's attentiveness to a photograph. For example, Peter continuously exclaimed, "Hey, that's me," while bringing a photograph close to his face for further examination. Additionally, there are also logistical benefits to using wearable cameras to film the bookmaking process as it frees the researcher's hands from holding a camera, allowing for a more authentic interaction between an adult and child without the awkwardness of a camera in between them (Robson, 2011).

Sensory Tours

Jacob: *I found lots of X-marks-the-spots and I saw some. I used GoPro [Sensory] Tours. I was looking for things with a GoPro on me.*

In the *Welcome too Our Forist* project, children were eager to wear cameras while playing and exploring in the forest. Once placed on their forehead, they seemed to forget that they were wearing them. Some even assumed the role of an investigator as illustrated in the opening quote by 4-year-old Jacob who showed pride in wearing the camera, telling each person that he happened upon in the forest, "I am wearing a GoPro. *Did you know whatever I look at is what records on this?*" He leaned close to "*take pictures*" of bugs on trees and other environmental features he found interesting during his forest exploration. As well, the recording of Jacob's interactions with peers and teachers provided insight into his EID.

The Sensory Tour method is informed by the tradition of walking tours used in geographical and environmental education research involving children to understand children's experiences and perspectives of place and their environment (e.g., Hart, 1979; Green, 2011, 2012; Sobel, 2002). Given the tendency of children to "talk while they are doing" (Parkinson, 2001, p. 145), walking tours provide an alternative format to engage children in conversations. Tours provide children with an opportunity to show and tell about what is important to them within a familiar and naturalistic setting (Hart, 1979; Green, 2012). In other words, "tours allow opportunities for children to show something that cannot be explained" apart from the setting (Green, 2012, p. 275). I have found that incorporation of a wearable camera enriches the walking tour method. Specifically, the wearable camera records a view of the world through the perspective of the person, in this case a child, who is wearing it (Chaflen, 2015).

In exploring children's EID, researchers can invite a child to put on a small camera during any environmental activity, whether it be picking up litter, climbing a tree, or exploring geographical features. While a small durable camera such as a GoPro is ideal because it is protected inside a plastic casing and can be worn around the forehead or chest, there are many other wearable cameras on the market that might be just as suitable. Perhaps, the main advantage of the Sensory Tour method is that it reveals a child's sensory interactions with his or her environment. The wearable camera captures what children see, hear, say, and touch. In viewing footage captured by a Sensory Tour, I can almost feel as if I am walking in the shoes of the child.

The wearable camera distinguishes when children turn their head or look up or down, when they run or fall, and the height and depth of their engagement with an environment. It also captures words exchanged with others, sounds and songs uttered under a child's breath, and their self-talk and expressions.

"*It's me here on this here … tree … its me!*" Oliver shouted from the top of a birch tree. In the *Welcome too Our Forist* project, a wearable camera revealed Oliver's close-up view of the tree and his heightened outlook of the forest landscape. The camera follows Oliver's gaze when he looked down and took notice of the skinny, fragile branch that he stood upon and uttered, "*Whoa!*"

A second advantage of the Sensory Tour method is that it provides adult researchers with a deeper insight of children's experiences of "being-in-the-world," understood as being in a particular space at a particular time (Heidegger, 1962). The method puts size into perspective by allowing the researchers to see how big adults and other environmental features appear to a child; it also makes apparent the features of the environment that a child attends to that might go unnoticed or be taken for granted by an adult. For example, in the *Welcome too Our Forist* project it became apparent how challenging and overbearing wild rosebushes can be to young children. The bushes that stand knee high to adults towered over the children. In this way, Sensory Tours provide a lens for researchers to understand the outer and inner dilemmas that children navigate in the construction of their environmental identity.

Thirdly, the Sensory Tours method also provides an unobtrusive means to explore children's experiences of an environment without the interference of an adult. In other words, wearable cameras provide an opportunity for children to engage in authentic peer-to-peer interactions undisturbed, removing the need for an adult with a video camera to loom nearby, prodding children's interactions. I found that some children who were quiet and reserved during group discussions were outspoken and leaders among their peers while exploring and playing in nature. Additionally, an adult's presence might interrupt the flow of children's make believe play. For example, in the *Welcome too Our Forist* project, the video of Priscilla's Sensory Tour revealed how a teacher's presence indirectly influenced her play. As Priscilla and Heather gathered leaves and pretended to make tea, the girls incorporated their actual physical desire for a drink into their play. When they noticed a teacher nearby, they dared one another to ask the teacher for a drink. Priscilla stated, "*We can't do that,*" indicating that asking for a drink is not something that they were allowed to do. Their non-verbal interaction also indicated that they were somewhat uncomfortable with the teacher's presence; the children's eyes got big and they postponed their play until they noticed the teacher was no longer heading in their direction. In this way, the Sensory Tour method captured otherwise unobservable interactions between two children, revealing insight into how peer-to-peer exchanges might inform a child's EID.

A fourth advantage of the Sensory Tour method is that the video captured is more likely to reveal more of the story of children's experiences in their environments than traditional video recording methods that typically rely on adults for positioning, and repositioning, the direction of the camera. Often adult researchers are only able to capture fragments and segments of participants' experiences, those that they happened upon or that caught their particular interest. In contrast, Sensory Tours go where a child goes and sees what a child sees. In this way, the footage can tell how many times a child visited a certain place or what events inspired such a visit. For example, in Heather's Sensory Tour, the entire video footage provided insight into her need for *Spatial Autonomy*, revealing her strong attachment to a tree that she claimed as her house in the forest. While the beginning of the video showed Heather and Priscilla preparing "*food*" in their "*house*," the middle of the footage revealed how Priscilla lost interest in Heather's "*house*," expressing her desire to explore other parts of the forest. Although Heather was reluctant, she went along with her friend only to express a desire and return to her house time and time again.

Lastly, an important benefit of Sensory Tours is that they are, for the most part, nonintrusive and enjoyable for children. The children in the *Welcome too Our Forist* project expressed that they liked wearing the cameras, indicating that it was one of their favorite activities of the entire project. Although wearing a camera was completely voluntary, children consistently volunteered over and over again. We have found that it works best if tours are structured as open-ended, lasting as long as the child wearing the camera is interested. If a child decides they no longer want to wear a camera, they can simply find an adult and ask that the camera be removed. Certainly in the *Welcome too Our Forist* project, we found that Sensory Tours that were less structured were better. At first we grouped children to collectively explore a topic. However, we found that although children were grouped according to common interests, they did not necessarily share the same experiences of those interests. For example, Rebecca and Oliver were grouped together to explore a common interest in "*castles*" in the forest. However, Rebecca's experience had been that of a *princess* castle, while Oliver was keenly interested in a *monster* castle. For this reason, placing Rebecca with Oliver for a collaborative Sensory Tour was not an ideal fit. While Rebecca tagged along during the beginning of the tour, she soon lost interest, which may have been related to Oliver's play script of a monster castle involving dragons and fights versus her own castle play script, which involved tea parties and princess fairies.

Rough play with the cameras may also pose challenges for researchers attempting to facilitate the Sensory Tour method. Rough housing with the wearable cameras was not uncommon during the *Welcome too Our Forist* project. Some children decided to play Camera *"Bull Fights,"* intentionally banging the cameras worn around their forehead against each other. The cameras were also sometimes hit with sticks or sideswiped by branches. Despite the wear and tear, the small cameras, housed inside tiny durable cases, sustained children's rough play without damage for the most part. Thus, I advise researchers to be selective about the type of wearable camera chosen for use, considering both the environmental context in which it will be used and the age and maturity of the children or adult users. I also recommend providing some basic guidance and directions to children to ensure proper treatment of the equipment at the beginning and throughout the project.

Finally, researchers should be sure to check the video size setting of the camera. Specifically, the default on some cameras might be set to a fairly large video size, and if it is not properly adjusted, can easily fill up the storage space on a computer in just a matter of days. Furthermore, once strapped on a child, be sure to check periodically to ensure that the camera is positioned at the correct angle and that it is properly recording. During the *Welcome too Our Forist* project, we found that every once in awhile a camera would inadvertently be shut off or the settings would be changed accidently.

Video-Stimulated Group Discussions

While the Sensory Tour method provides an opportunity for children to engage in data collection, video-stimulated recall discussions can be implemented to engage children in analysis and interpretation of the video that they collect. Video-stimulated recall has become increasingly popular in participatory research involving young children (Thomson, 2008). The method "involves video-recording an activity and then replaying the recording to the participants so that they can comment on matters of interest" (Rowe, 2009, p. 427). Foreman (1999) suggests that replaying videos for children can serve as a "tool of the mind" (p. 1), inviting children to interpret the meaning of their actions. Video images can also stimulate aesthetic and emotional responses (Thomson, 2008). Researchers have paired video-stimulated recall with semi-structured interviews (Rowe, 2009), and reflective dialogues between children and teachers (Robson, 2011). Using the video-stimulated recall for group discussions can promote intersubjectivity and collaborative reflection among children (Dahlberg, Moss, & Pence, 2007). Rowe (2009) suggested that engaging participants in video-stimulated recall can provide an "insider's perspective" (p. 434) on actions, behaviors, and experiences, and provide participants with opportunities to raise ideas that have not been previously thought of by a researcher.

In the *Welcome too Our Forist* project, video-stimulated recall was used in the context of group discussions to engage children in analyzing their experiences in the forest and for selecting their own research topics. The children collected video footage of their first few days in the forest using Sensory Tours. Afterwards, the research team reviewed the video footage and identified short segments of children's activities that might be of particular interest to the children and that could be used to stimulate discussions. The video was played to small groups of children in their classroom and the following questions were posed: "What were you thinking about when you watched the movie? What did you notice? What did you hear? What did you see? What are you wondering about?" The children were also invited to draw pictures to represent and tell about their experiences. Their discussion comments and drawings were incorporated into the children's Research Big Book.

Children were excited about seeing themselves in the videos, pointing and saying, "*Hey that's me!*" or "*Where's me?*" They also identified peers who they recognized in the videos. The video as well as the prompted discussions encouraged children to explore their natural world socialization:

Researcher:	What did you see and think about in that movie?
Peter:	*I wonder why the monsters have bugs?*
Researcher:	Okay, What about the kids and the bugs? Do you think about anything with the kids and the bugs?
Cindy:	*You smash ladybugs?*
Researcher:	You want to know if kids smash ladybugs?
Cindy:	*Yeah.*
Researcher:	Okay, do kids smash ladybugs?
Cindy:	*And then people say … you don't smash ladybugs.*
Researcher:	They do?

While Peter's comments seemed to reveal more about his *Environmental Competency*, that is, his imaginary encounter with bugs and monsters, Cindy's comments alluded to a tension between *Environmental* Action and *Environmental Harm*, namely should kids preserve or smash bugs in the environment?

In addition to bugs, the children also expressed interest in other topics, including: rosebushes, sticks, "*X-marks-the-spots*," "*forts*," "*castles*," and "*houses*" claimed in the forest. For instance, Nathan shared, "*Me, Paul and Heath, and Garrett and me, and we investigating them and we also found a little fort and a little tree. And we kind of like it.*" Nathan's statement revealed a shared quest for *Spatial Autonomy*. His comment, "*we kind of like it,*" indicates his preference for his special place in the forest.

Artistic Representations

"Like language, art is a symbol system that can be used to generate meaning" (Isenberg & Jalongo, 2001, p. 106). Art as a method has grown in popularity in childhood environmental education research. Indeed, in a recent review of research methods in early childhood environmental education, over half, i.e., 19 of the 36, studies evaluated incorporated some type of artistic activity (Green, 2015). Such engaging and interactive methods provide children with a comfortable and creative space to express their understandings and perspectives on a topic. Children's drawings, writings, maps, and photographs are some of the most popular forms of artwork that have been adopted in environmental education research. However, for very young children who are still developing their fine motor skills, abstract forms of artwork may be more developmentally appropriate, including painting, molding with play dough, building with blocks, and role-play. In promoting children's agency in research, it is important to provide children with several options for representing their ideas.

In the *Welcome too Our Forist* project, the children created and interpreted artistic data in a variety of forms. Three centers (Drawing and Painting, Role-Playing, and Building a Model) were constructed in the forest for children to represent their thoughts, perspectives, and ideas about their environmental experiences. Initially, a rotation schedule was devised to enable groups of children to circulate through the centers over the course of four days. However, after using the schedule for a couple of days, we recognized that assigned centers did not necessarily support children's agency in the research. Children who were painting and drawing wanted to role-play and children building models wanted to paint or draw (or vice versa). Thus, we relinquished the use of a schedule and invited children to freely explore whichever center most interested them. The only drawback of encouraging free choice of centers was that sometimes centers became overly populated or children were unable to complete their projects before it was time to leave the forest.

Drawing and Painting

Children can create paintings and drawings to represent their experiences of nature. As well inviting children to describe their artwork "record[s] the journey of their constructions of meaning" (Einarsdóttir et al., 2009, p. 219). In this way, the focus is placed on the meaning making process rather than children's artistic abilities or their finished products. In the *Welcome too Our Forist* project, we constructed a Drawing and Painting center by placing an 8X6 foot board on four tree stumps for a table, stringing a line between two trees for a drying station, and supplying paint, paper, markers, and cardboard. A sense of excitement buzzed around the art table, as children painted and drew about bugs, X-marks-the-spots, various places, and activities they liked to do in the forest. During one art center session, a group of children interacted with a caterpillar that happened to cross Heidi's paper.

Derek: *Hey there's a caterpillar on your …*

Heidi drops her marker in the bucket and looks up at Ms. Bethany, pointing to the caterpillar.

Heidi: *Ms. Bethany, there's a caterpillar on my page.*
Ms. Bethany: Oh wow.
Ms. Taylor: Did you draw a cater … oh is that …
Heidi: *No, it's a real one.*
Ms. Taylor: Oh, it's a real caterpillar right there.
Ms. Bethany: He wants to see your art there, Heidi.

Heidi touches her paper near the caterpillar and laughs.

Heidi: *I think he wants … I think … I think he thinks it's a real tree.*

Creating art in the forest allowed for unique interactions with forest creatures that could not be had indoors. During the interaction with the caterpillar, Heidi along with her teachers personified the caterpillar, making assumptions about its intentions and behavior. As an individual's environmental identity is formed through interactions and experiences in nature, Heidi's empathetic response towards the caterpillar is indicative of her growing relationship with the natural world. In considering Heidi's EID, or natural world socialization, we can see how the environment is influencing her development as well as how Heidi is actively shaping her environment.

While the literature points at the effectiveness of children's drawings as a research method (Clark, 2005; Einarsdóttir, 2007; Einarsdóttir et al., 2009; Punch, 2002), painting can provide an equal, if not more effective, means for engaging younger children who are still developing their fine motor skills (Green, 2013). Additionally, inviting children to choose among various art forms accounts for their personal preferences and range of abilities.

Setting up opportunities for socializing through group art making encourages children to engage in discussions about what is meaningful to them, communicate shared experiences, and bounce ideas off of one another. In this way, the art table creates a common ground for children to express their particular interests in topics. That being said, social influence can also pose limitations in that children can be easily influenced by the activities of others around them and might strive to create similar artwork (Einarsdóttir et al., 2009). Furthermore, children might try to construct what they perceive the researcher wants, rather than authentically express their own ideas and perceptions. One way to mitigate social influence is to include multiple methods for children to share their perspectives of their environment. This also allows for triangulation of the various data sources. It is also important to keep the size of groups manageable. Hosting too many children at an art center may make it challenging for researchers to capture descriptions of artwork. If large groups are facilitated, researchers should ensure plenty of help for hanging up finished pictures and recording children's interpretations of their artwork.

In considering possible limitations of art making as a method, some children might create artwork that seems unrelated to the environmental context. It is important, however, not to dismiss what a child creates as irrelevant; rather a researcher might probe further to ask the child to explain the connection that she is making to a particular setting. Furthermore, there are logistical considerations when facilitating art activities with children especially in nature; spill proof paint containers and non-toxic products support the common goal of consideration for and preservation of the environment.

Building a Model

Building and molding is a common feature of childhood play, whether indoors or outdoors, in formal (classrooms) or informal (forest) settings, children love to explore how "stuff" fits together and what that "stuff" might become. Indeed, "children learn best through manipulation of materials in which they can see the effects they have on the world around them" (Swartz, 2005, p. 100). In other words, through building and molding, children are interpreting and constructing their own sense of place in their environments. Childhood place research has revealed children's inclination to manipulate "loose parts" or objects in nature (Hart, 1979; Kjørholt, 2003; Kylin, 2003; Sobel, 2002). Cobb (1977) described this as "a sort of fingering over the environment in sensory terms, a questioning of the power of materials as a preliminary to the creation of a higher organization of meaning" (p. 48). In other words, by manipulating objects and settings through their play, children are personally making meaning of the world around them and developing their sense of environmental identity. Extending to a research context, I propose Building a Model as an interactive and engaging method for exploring children's EID.

In the *Welcome too Our Forist* project, we invited children to use natural materials to build models to represent their environmental experiences. A Building a Model center was established in the forest with a tree stump table, buckets for children to gather natural materials, play dough, glue, markers, shoeboxes, and other pieces of cardboard for model construction. First, children explored and gathered materials from their environment, including sticks, leaves, spruce pine, bark, moss, and mushrooms. Next, children shaped and glued materials onto cardboard surfaces, creating miniature worlds to represent both real and imaginary elements of their experiences. After children were done constructing their models they were invited to describe them. Bug homes, mini villages, tiger huts, tiny lakes, and rivers were among the features depicted in children's models, some of which accurately represented the local flora and fauna of their environment and some of which extended features to include non-native species that were not part of their bioregion (e.g., tigers).

The primary advantage of the Building a Model method is that models constructed by children represent their particular interests in their environment. For instance, Sergo built his home in the forest, which validated his ongoing quest for *Spatial Autonomy*. Katherine built a ladybug on a tree, revealing her *Environmental Competency* in knowing the creatures that she had interacted with in the forest. Others incorporated native fauna species that made up the local habitat even if they had not come into direct contact with them during class forest explorations (e.g., bears, salmon, and berries). Nevertheless, such depictions provided insight into their personal interactions with the local outdoor environment, both past and present.

Additionally, children's descriptions of their models can provide understanding of the social, cultural, geographical, and familial influences on their EID. Several children created their houses and described features of their yard (e.g., kid pool). They also talked about going on hikes with a family member, or engaging in hunting and gathering activities like berry picking. These all make up the local ethos of their place (Kjørholt, 2007).

As well a third advantage, which could also be a disadvantage of the Building a Model method depending on how you view it, is that children incorporate both real and imaginary elements in their models in order to interpret their surroundings. While this can be viewed as an advantage, especially in considering fantasy play as an important feature of EID, it might also be perceived as a disadvantage in that their representations may not fully reveal the truth of their previous experiences. Facilitating multiple methods to explore children's experiences, as discussed earlier, provides one way to overcome this. If children represent and tell about an experience or particular aspect of their environment over and over again through multiple methods then this is likely a salient aspect of their EID.

Additionally, other disadvantages or challenges associated with Building a Model is that this method may not be suitable for especially young children who are still developing their fine motor skills, so it is important to keep in mind the age and skill level of children when considering this method. For example, it may be challenging for some children to use liquid glue or to cut or shape materials into distinct environmental features. Thus, when facilitating this method with young children, researchers may want to involve an ample number of adult or peer assistants. It is also wise to arrange for enough assistance with video or audio recording children's descriptions of their models. Because models may be very abstract and indistinguishable, children's explanation of what they created is of upmost importance. Finally, allow plenty of time for children to create and construct their models. Rushing the process may lead to gaps or a disjointed understanding of a child's EID.

Role-Playing

Childhood is full of fantasy play and creativity. Why not harness children's imagination by employing role-play as a data collection method? It requires children to consider ideas from various perspectives and draw upon their own beliefs, values, and experiences. Thus, role-playing is a useful data collection method for studying children's EID. As O'Sullivan (2011) explains:

> Role-play is concerned with representing and exploring different people's points of view, and different points of view forge different types of knowledge. It places participants at the centre of the learning experience, and allows them to build their own bridge of understanding. As a result of this informed consideration, they are better able to resolve problems and issues. (p. 513)

By assuming the role of human or a member of the more-than-human world (plants, animals, or other environmental features), children begin to explore their personal relationship with nature. Additionally, through role-playing children engage in empathetic reasoning, that is, they think about how it might feel to be someone or something else, like a bug, sunflower, or another type of creature (Donohoe & O'Sullivan, 2015). Additionally, children's *Environmental Competency* is harnessed through perspective-taking and emotional understanding. This encourages children to think about how they feel about and relate with various entities in nature. Furthermore, role-playing promotes children's social engagement with peers, and when orchestrated outdoors, children are more likely to relate their stories to their environment. In other words, they are more likely to incorporate elements of the natural world into their role-play. They may use a stick as their sword or build a nest out of pinecones.

In the *Welcome too Our Forist* project, a stage was built for the Role-Playing center, with a sheet strung between two trees, a tarp laid across the forest floor, with puppets of a flower, tree, and sun, ladybug and butterfly costumes, and other prompts like a talking stick made available for children to act out their forest experiences. First, children were invited to choose the costume or puppet that made them feel most comfortable. Their sense of comfort supported their feelings of *Trust in Nature* and promoted their engagement as active researchers. The researcher also assumed a pretend role, entering into play with the children. In this way, the researcher engaged as a facilitator as well as an active participant of the experience. Next, children were reminded of the topics in which they had previously expressed an interest (i.e. bugs, X-marks-the-spot, sticks, forts, houses, castles). They were then encouraged to construct a narrative based on their role selection and their expressed interest in a topic. Question prompts were used to encourage children to think about and facilitate their participation in the role-play: "Let's think about what we like to do in the forest. How can we make a story about what we like to do? Can we use things around us to make that story? What might you say? What might you do?"

Children responded by assuming different aspects of their environment and reenacting key interactions of living flora and fauna of the forest through their perspective. In this way, they demonstrated their understanding of environmental relationships and how humans interact with particular features of their environment. Additionally, the researcher participated by assuming a role, often one that was assigned by children. This further promoted children's agency in the role-play process. For instance, in one scenario the children asked the researcher to play the role of the "Forest," in another situation the researcher was invited to be "Mr. Stickerbush." Puppets or costumes were used to help act out the given roles. By taking on the role of an environmental feature, the researcher was able to evoke the children's experiences and gain deeper insight into how they perceived their own experiences within the forest. As the "Forest," the researcher was able to ask the children how they interacted with it, what they liked best about it, and what the forest meant to them. As "Mr. Stickerbush," the researcher asked similar questions and the children actively responded by lightly stroking the "Mr. Stickerbush" puppet and then pretending to be poked by a wild rosebush in real life.

Studies have shown that discourse between children and an adult researcher is enriched when researchers assume a role compared to discussions generated by a traditional researcher (Aitken, 2014). In turn, children may feel more comfortable with researchers engaged in playing a role; this can "open up possibilities for new storylines and admissible actions" (Aitken, 2014, p. 255). Similarly, research has also shown that children are more likely to view puppets as peers rather than as authority figures (Belohlawek, Keogh, & Naylor, 2010; Simon, Naylor, Keogh, Maloney, & Downing, 2008). We found this to be the case, including with shy children who were more inclined to engage in conversations with puppets than with adults (Luckenbill, 2011; Keogh & Naylor, 2009). Thus, by taking on an imaginary role researchers are venturing towards new avenues for engaging children in expressing aspects of their environmental identities.

Another benefit of role-playing as a data collection method is that children perceive it as fun and engaging. In the *Welcome too Our Forist* project, children were drawn to the Role-Playing center, curious and eager to join in the excitement. Many revisited the center time and time again, enacting and reenacting similar stories on multiple occasions. This served to validate children's experiences and interests in particular topics. As well the benefit of role-playing as a research method is that it offers flexibility to cater storylines to fit both the shared ideas of groups as well as individual ideas on any given topic. For instance, while a group might decide to create a story about bugs in an environment, individual children can express how they think and feel about a particular bug.

While social dramatic play is a common activity among young children, role-playing as a research method is a bit more structured and focused. In planning to use this method, researchers should purposely think about which types of roles are appropriate for a given environment. To the extent possible, costumes and prompts should reflect the flora and fauna of that setting. It is also important for researchers to have materials on hand for children to design additional costumes to fit with the story lines that they conjure up. We found that storylines were often limited to the props available to children. Thus, by creating a stage in nature, children can be encouraged to incorporate elements from their environment to enhance their role-play.

Conclusion

In this chapter, several qualitative methods were presented for exploring children's EID. From bookmaking to role-playing, hands-on methods encourage children's active engagement in the research process. Methods should be designed with flexibility in mind, adaptable to children's desires and needs. What works well in one project may not work well in another or what works well with one group of children may not be fitting for the same group on a different occasion. Additionally, as children's environmental identity continues to develop and form, their interests can change from one day to the next. It is important, therefore, that researchers keep on their toes and remain open to the shifting nature of children's budding personalities and interests.

Six methods were presented in this chapter; each has its own advantages and challenges. A common advantage of each method is that they are child-led and directed. Bookmaking invites children to explore their ideas about the concept of research, choose their data collection methods, take an active role in choosing their research topics, and document their thoughts and beliefs through quotes, artwork, and photographs. Books can also be used as a means for children to present their findings of a research project to family, friends, and community members. Sensory Tours is a novel research method that allows children opportunities to video record their own experiences of an environment; it embodies the experience of a child by capturing what they say, see, and touch in nature (Green, 2016). In other words, a Sensory Tour shows what draws a child's interest, when they look up or down, and the height and depth of their environmental experiences. Video-stimulated recall discussions can be paired with Sensory Tours to invite children to analyze and interpret their own experiences. In this way, they are encouraged to revisit an environmental experience and reflect on the significance of their actions and interactions with peers and environmental features. Several forms of artistic expressions were presented including drawing and painting, building a model, and role-playing. These various art forms promote children's agency in sharing their perspective of their environment. And when facilitated in nature, children are encouraged to incorporate materials from the environment into their representations. When selecting methods, a researcher should consider the advantages and challenges of each of these methods and pair multiple methods in order to more fully explore children's EID.

There is a growing interest in environmental education in engaging children as active researchers (Barratt Hacking et al., 2013; Green, 2015). While on paper these methods may appear seamless and easy to implement, promoting child-led and child-driven research is anything but simple. It requires researchers to engage in reflexivity by constantly taking a step back and examining their own ideas and notions of what a project should entail. From encouraging a child to glue a picture on the "right way" to quickly dismissing a child statement because it is interpreted as "unrelated," researchers can intentionally or unintentionally disrupt children's agency in determining the course of a project. I recognize that it can be challenging as a researcher to relinquish control, even in the little things, but I am convinced that with more practice and with good intentions it can be done. And I argue that doing so is important if we really want to understand children's EID, we must allow for and promote children's agency of expression.

Note

1. The methodology and methods presented in this chapter were first published in the following article: Green, C. (2016). Four methods for engaging children as environmental education researchers. *The International Journal of Early Childhood Environmental Education*, 5(1), 6–20. Material from the article was adapted and republished with permission from the journal editor.

References

Aitken, V. (2014). From teacher-in-role to researcher-in-role: Possibilities for repositioning children through role-based strategies in classroom research. *Research in Drama Education*, 19(3), 255–271.

Barratt Hacking, E., Cutter-Mackenzie, A., & Barratt, R. (2013). Children as active researchers: The potential of environmental education research involving children. In R. B. Stevenson, M. Brody, J. Dillon, & A. E. J. Wals (Eds.), *International handbook of research on environmental education* (pp. 438–458). New York, NY: Routledge.

Belohlawek, J., Keogh, B., & Naylor, S. (2010). The PUPPETS project hits WA. *Teaching Science*, 56(1), 36–38.

Chaflen, R. (2014). Your panopticon or mine?' Incorporating wearable technology's Glass and GoPro into visual social science. *Visual Studies*, 29(3), 299–310.

Clark, A. (2005). Listening to and involving young children: A review of research and practice. *Early Child Development and Care, 175*(6), 489–505.

Cobb, E. (1977). *The ecology of imagination in childhood*. New York: Columbia University.

Conroy, H., & Harcourt, D. (2009). Informed agreement to participate: Beginning the partnership with children in research. *Early Child Development and Care, 179*(2), 157–165.

Corsaro, W. A. (2015). *The sociology of childhood* (4th ed.). Thousand Oaks, CA: Sage.

Dahlberg, G., Moss, P., & Pence, A. (2007). *Beyond quality in early childhood education and care: Languages of evaluation* (2nd ed.). New York, NY: Routledge.

Donohoe, P., & O'Sullivan, C. (2015). The bullying prevention pack: Fostering vocabulary and knowledge on the topic of bullying and prevention using role-play and discussion to reduce primary school bullying. *Scenario, 9*(4), 19–36.

Duhn, I. (2012). Making "place" for ecological sustainability in early childhood education. *Environmental Education Research, 18*(1), 19–29.

Einarsdóttir, J. (2007). Research with children: Methodological and ethical challenges. *European Early Childhood Education Research Journal, 15*(2), 197–211.

Einarsdóttir, J., Dockett, S., & Perry, B. (2009). Making meaning: Children's perspectives expressed through drawings. *Early Childhood Development and Care, 179*(2), 217–232.

Foreman, G. (1999). Instant video revisiting: The video camera as a "tool of the mind" for young children. *Early Childhood Research and Practice, 1*(2). Retrieved from http://ecrp.uiuc.edu/v1n2/forman. html.

Graham, A., & Fitzgerald, R. (2010). Progressing children's participation: Exploring the potential of a dialogical turn. *Childhood, 17*(3), 343–359.

Green, C. (2011). A place of my own: Exploring preschool children's special places in the home environment. *Children, Youth, and Environments, 21*(2), 118–144.

Green, C. (2012). Listening to children: Exploring intuitive strategies and interactive methods in a study of children's special places. *International Journal of Early Childhood, 44*(3), 269–285.

Green, C. (2013). A sense of autonomy in young children's special places. *International Journal of Early Childhood Environmental Education, 1*(1), 8–33.

Green, C. (2015). Towards young children as active researchers: A critical review of the methodologies and methods in early childhood environmental education research. *Journal of Environmental Education, 46*(4), 207–229.

Green, C. (2016). Sensory tours as a method for engaging children as active researchers: Exploring the use of wearable cameras in early childhood research. *International Journal of Early Childhood, 48*(3), 277–294.

Hart, R. (1979). *Children's experience of place*. New York, NY: Irvington.

Hart, R. A. (1997). *Children's participation: The theory and practice of involving young citizens in community development and environmental care*. New York, NY: Earthscan.

Haw, K. (2008). Voice and video: Seen, heard and listened to. In P. Thomson (Ed.), *Doing visual research with children and young people* (pp. 192–207). London, UK: Routledge.

Heidegger, M. (1962). *Being and time*. (trans: J. Macquarrie & E. Robinson). New York, NY: Harper & Row.

Isenberg, J. P., & Jalongo, M. R. (2001). *Creative expression and play in the early childhood* (3rd ed.). Columbus, OH: Merrill-Prentice Hall.

James, A. (2009). Agency. In J. Qvortrup, W. A. Corsaro, & M. S. Honig (Eds.), *The Palgrave handbook of childhood studies* (pp. 34–45). New York, NY: Palgrave MacMillan.

James, A., & Prout, A. (1990). *Constructing and reconstructing childhood: Contemporary issues in the sociological study of childhood.* Bristol, PA: Falmer Press.

Ji, O., & Stuhmcke, S. (2014). The project approach in early childhood education for sustainability: Exemplars from Korea and Australia. In J. Davis & S. Elliott (Eds.), *Research in early childhood education for sustainability: International perspectives and provocations* (pp. 158–179). New York, NY: Routledge.

Keogh, B., & Naylor, S. (2009). Puppets count. *Mathematics Teaching, 213,* 32–34.

Kjørholt, A. T. (2003). Creating a place to belong: Girls' and boys' hut-building as a site for understanding discourses on childhood and generational relations in a Norwegian community. *Children's Geographies, 1*(1), 261–279.

Kjørholt, A. T. (2007). Childhood as a symbolic space: Searching for authentic voices in the era of globalization. *Children's Geographies, 5*(1/2), 29–42.

Kylin, M. (2003). Children's dens. *Children Youth and Environments, 13*(1), 30–55.

Lee, N. (2001). *Childhood and society: Growing up in an age of uncertainty.* Buckingham: Open University Press.

Lincoln, Y. S., Lynham, S. A., & Guba, E. G. (2011). Paradigmatic controversies, contradictions, and emerging confluences, revisited. In N. Denzin & Y. S. Lincoln (Eds.), *The Sage handbook of qualitative research* (4th ed.) (pp. 97–128). Thousand Oaks, CA: Sage.

Luckenbill, J. (2011). Circle time puppets: Teaching social skills. *Teaching Young Children, 4*(4), 8–10.

Mackey, G. (2014). Valuing agency in young children: Teachers rising to the challenge of sustainability in the Aotearoa New Zealand early childhood context. In J. Davis & S. Elliott (Eds.), *Research in early childhood education for sustainability: International perspectives and provocations* (pp. 180–193). New York, NY: Routledge.

O'Sullivan, C. (2011). Role-playing. In L. Cohen, M. Lawrence, & K. Morrison (Eds.), *Research methods in education* (7th ed.) (pp. 510–522). New York, NY: Routledge.

Parkinson, D. D. (2001). Securing trustworthy data from an interview situation with young children: Six integrated interview strategies. *Child Study Journal, 31*(3), 137–156.

Phillips, L. G. (2014). I want to do real things: Explorations of children's active community participation. In J. Davis & S. Elliott (Eds.), *Research in early childhood education for sustainability: International perspectives and provocations* (pp. 194–207). New York, NY: Routledge.

Punch, S. (2002). Research with children: The same or different from research with adults? *Childhood, 9*(3), 321–341.

Robson, S. (2011). Producing and using video data in the early years: Ethical questions and practical consequences in research with young children. *Children and Society, 25*(3), 179–189.

Rowe, V. C. (2009). Using video-stimulated recall as a basis for interviews: Some experiences from the field. *Music Education Research, 11*(4), 425–437.

Schutz, A. (1967). *The phenomenology of the social world*. Evanston, IL: Northwestern University Press.

Schwandt, T. A. (2015). *The Sage dictionary of qualitative inquiry*. Thousand Oaks, CA: Sage.

Simon, S., Naylor, S., Keogh, B., Maloney, J., & Downing, B. (2008). Puppets promoting engagement and talk in science. *International Journal of Science Education, 30*(9), 1229–1248.

Sobel, D. (2002). *Children's special places: Exploring the role of forts dens, and bush houses in middle childhood*. Detroit, MI: Wayne State University.

Swartz, M. I. (2005). Playdough: What's standard about it? *Young Children, 60*(2), 100–109.

Taylor, N. J., & Smith, A. B. (Eds.). (2009). *Children as citizens? International voices*. Dunedin, New Zealand: Otago University Press.

Thomson, P. (2008). Children and young people: Voices in visual research. In P. Thomson (Ed.), *Doing visual research with children and young people* (pp. 1–19). London, UK: Routledge.

United Nations. (1989). *Convention for the rights of the child*. New York, NY: Author.

United Nations. (2005). *Convention on the rights of the child: General Comment No. 7. Implementing child rights in early childhood*. Geneva, Switzerland: Author.

· 7 ·

DIVERSE OBSERVATIONS OF EID

Voices from the Field

Vivid examples of children's EID have been woven throughout each of the previous six chapters of this book. These examples primarily stemmed from my own research with young children as well as personal and educational memoirs. This chapter adds more depth to the discussion by interweaving the voices of three other Alaskan educators who have or are contemplating how they might apply EID in their own work with children and youth from Alaska, Finland, and Thailand. The three authors reflect on the very personal nature of EID, and it's overlap with culture, geography, and place. Their different orientations are not presented as an argument towards one or another "ideal" orientation of EID, rather each orientation is recognized as a reflection of the reality of the diverse world in which we live in. Early experiences and opportunities shape us, how we come to view our lives, our purpose, and what we consider important. These experiences are interwoven, though they appear distinct and separate, we live in a world that is interconnected—what humans do in one part of the globe inevitably impacts the other.

In the first essay, Karen Martin, a middle school teacher from a small community at the base of Denali National Park in interior Alaska writes about the importance of cultivating global connections. She reflects on an educational partnership that was forged between her students in Alaska and students from Finland. By creating opportunities for cross-cultural dialogue about climate change and sustainability, Martin hopes to enhance her student's *Environmental Competency* and promote *Environmental Action*. Further, by exposing students to rich meaningful encounters in nature and encouraging them to work through challenges, Martin hopes to shift student's environmental identity, from what she observes as an anthropocentric orientation to one that recognizes complex inter-relations.

Similarly, the second essay by educator Robin Child illustrates two unique pathways of EID. She compares and contrasts the lives of her two ten and eleven-year-old cousins, one from Thailand and the other near Anchorage, Alaska. While both have been heavily exposed to nature since infancy, the cultural forms in which such experiences have taken root have differed. Child outlines the essential components of how *Trust in Nature* is nurtured by parents and caring adults, forming the foundation of *Spatial Autonomy* and *Environmental Competency*. In considering *Environmental Action*, Child notes the two boys' perceptions of environmental concerns are of a different magnitude and specific to the place in which each is being raised. This adds an important question about the end goal of EID: how should children be nurtured in their EID to address pervasive environmental concerns? How, if at all, should such nurturing differ depending on specific issues or concerns?

In the third reflection, Angela Lunda takes us on a journey from her childhood at fish camp, deeply immersed in her Tlingít culture. She parallels EID with cultural identity development, highlighting the importance of make-believe play in nature, storytelling, family collaboration, and a strong work ethic. Lunda's essay, similar to Martin and Child's, reveals the unique pathway of EID in relation to culture and place. Unlike the other two reflections, however, Lunda's experiences were grounded in Indigenous cultural subsistence practices and a deeply rooted understanding of the reciprocal relations between humans and other living beings. Lunda closes by posing suggestions for strengthening EID and culturally relevant pedagogy among teachers and their students in Alaska and beyond.

Transferring Passion to Arctic Conversations About Climate Change

By Karen Martin, University of Alaska Fairbanks

Introduction

How can I help my rural Alaskan middle and high school students develop a passion for environmental sustainability and climate change mitigation? This question speaks to the heart of my work as a teacher in rural Alaska. Like most of my students, I grew up with many opportunities to play outside and explore the natural world. During my adolescent years, I lived in Shoshone National Forest in northern Idaho. I spent entire summers from dawn until dusk either on the banks of the St. Joe River or completely immersed in it. As a young adult, one of my most formative jobs was as a park ranger at Mount Rainier National Park, where I had the opportunity to facilitate visitor experiences at one of my most cherished places on earth. Later I moved to Alaska with my husband to further explore the rivers, glaciers, and ice fields in the far north. My mother, who imparted her love of the outdoors to all of her children, directly nurtured the development of my own sense of *Trust in Nature*. I also attribute my strong sense of *Environmental Competency* to this trust that guided me in acquiring skills to read rivers, climb frozen waterfalls, and traverse crevassed glaciers. The mountains have always been a dependable source of constant and enduring beauty and strength to me when other sources of comfort failed. Early in my life, I developed an innate sense of responsibility for my actions and how they impact the rest of the natural world.

In order to nurture learning experiences that move my students toward a passion for environmental sustainability and addressing issues surrounding climate change, I received a *Fund for Teacher Fellowship* to visit Finland and collaborate with Finnish teachers to co-create learning curriculum for middle and high school students, ages 12–17 years old (fund for teachers, 2013). The curriculum focuses on collaborative learning, conversation about climate change, and students' sense of connection to nature. Learning experiences also utilize citizen science methods such as *The GLOBE Program: Global Learning and Observations to Benefit the Environment* (NASA, 2018). In this essay, I describe the scope of this project in relationship to EID.

Project Context

I teach in Healy, Alaska, located on the Parks Highway that connects Fairbanks to Anchorage. The community of Healy is two hours from any major service such as a hospital, grocery stores, university, and other medical services. Unlike many communities in rural Alaska, Healy is accessible by road and has a relatively high median family income. It is predominately White (94%), with only 2% representing Alaska Native groups and 4% Asian (U.S. Census, 2015). The major economic providers of jobs are Usibelli Coal Mine, Denali National Park, Denali Borough School District, construction and transportation services such as Alaska Department of Transportation, and other tourism related entities such as Holland America Cruise Lines.

Our project involves connecting my middle and high school students with peers in Finland to engage in collaborative inquiry and conversations about climate change. The subject of climate change is a controversial topic in our community and deeply rooted in threats to the livelihood of my students' families. A majority of the families in our community work at Usibelli Coal Mine, which has experienced a reduction in international and domestic orders for coal in recent years as a result of global actions to reduce carbon emissions. Students in our classroom are aware of their families' positions on climate change influencing their jobs. Additionally, students in our community are geographically and culturally isolated; as such students seem to be somewhat disconnected from recognizing how environmental degradation that occurs in one part of the globe inevitably impacts another.

Initial Observations

In my initial observations of the project, I noted that my students seem to have what could be described as apathy toward environmental advocacy. Despite their knowledge that climate change is happening, I believe that my students lack an emotional connection to the issue. Students have identified this position in their own words in our classroom. For example, students have stated that other issues such as bullying are more important to them than climate change mitigations as well as shared doubts whether local actions such as recycling or carpooling can really make a difference. These ideas from students represent elements of *Environmental Disdain*, despair, and even apathy, which I see as hallmarks of indifference.

Through an open-ended concept map exercise, my students identified elements of the natural world such as mountains, wildlife, cold, weather, and

Denali National Park as critical elements of their local context in response to the question, "What is my side of the world like?" In letters they wrote to their Finnish peers, my students independently chose to write about elements of our community context that included specific natural attributes such as the length of day, darkness, snowfall, animals, and the aesthetic beauty of the mountains. When introducing themselves through digital media, including videos made through iMovies and electronic posters, their words and images included outdoor activities with their families, sports activities, and favorite hobbies such as video games, reading, drawing, and dancing. Their choices reflect connections to both the natural and material world as important aspects of their identities.

These findings are supported by my own observations of watching these students grow up in our community for twelve years. I taught most of these students and their siblings in kindergarten and first and second grades. All of my students' families have engaged their children in a range of experiences with the natural world since they were very young. Activities include extensive hunting experiences, camping, backpacking, hiking, climbing, berry picking, ice skating, skiing, swimming, river trips, horseback trips, outdoor science and art explorations, and learning from naturalists and scientists. Several of my students have lived in cabins without running water where their experience with nature involves basic survival activities such hauling water from local streams and going to the bathroom outside.

It is important to acknowledge, however, that our rural Alaskan context is somewhat unique to Alaska in that demographically my students are not Alaskan Native or of Indigenous ethnicity. Indigenous Alaskans have a distinctly different relationship to the environment through a strong subsistence lifestyle that is fundamentally based on beliefs about the interconnectedness of all living things (Kawagley, 2006). This mindset is foundationally oriented towards sustainability, living in balance with nature, and it is also more community-oriented (Kawagley, 2006). The connection that most of my own students experience with the natural world is not founded on a subsistence-based epistemology, and the environmental activities my students engage in are not uniformly positioned with respect to valuing all living beings as in Alaskan Native culture. Rather most of my students' environmental identities seem to be framed around an anthropocentric mindset, centered predominately on the benefits of the environment for human beings and less on the influence or impact of their behaviors and actions on an interconnected ecosystem. I believe this poses the central challenge to promoting *Environmental Action* with my students, especially in regards to climate change. Thus, fundamentally, the opportunity to collaborate internationally with another group of students is grounded in my desire to create opportunities for my students to think more holistically about their relationship with the environment and the implications of their actions.

Another observation relevant to this project and important to present is that our students' connection to the environment is also influenced by their involvement in indoor, extracurricular activities as well as the economic dependencies of their families. The majority of my students are engaged in extracurricular sports programs from September to May that require a commitment to practice sports almost two hours after school each day and long trips nearly every weekend. Additionally, many students come from families whose parents work at the local coal mine, which plays a major role in sustaining the economy of the community. Taken together, these realities strongly influence perceptions about why my students live in their community and the types of relationships they form with their environment. There is a strong pull to a material culture and disconnect from environmental advocacy.

Reflexive Repositioning and Initial Findings

Careful consideration of EID has caused me to reflect on and become aware of my own interactions and intentions in my EE design work. I realize that I have been placing on students my own imperative to fix climate change, which is a direct result of my own sense of *Environmental Action*. This reflection has caused me to think more deeply about the environmental experiences of my students and their relationships to their environment. The initial goals of my project design were aimed at developing knowledge about climate science and identifying the cultural influences in which my students are immersed. I realized, however, that I have not recognized that perhaps my students' emotional connection with the environment is not as strong as I assumed. The EID model thus presents a theoretical framework that is important for me to consider. Specifically, in order to help guide my students' progress toward *Environmental Action*, I must first support the affective aspects of *Environmental Competency* by recognizing and creating opportunities that provoke strong emotional ties and a reconnection to the environment.

Possibilities for Future Design Work Influenced by EID Theory

EID presents a critical emphasis on understanding the mediation of action through feelings and emotions about the environment. In chapter one, Carie shares how EE approaches have been placing too much emphasis on teaching about environmental problems, resulting in children developing a sense of helplessness or fear of nature. Sobel (1996) also argued that global issues were being placed on children's shoulders much too early. Instead, "children [should] have the opportunity to bond with the natural world, to learn to love it and feel comfortable in it, before being asked to heal its wounds" (p. 13). It is clear in the EID model that emotional connection to nature plays a significant role in shaping environmental values and behaviors that influence children's agency and actions. Relevant to my design work and intentions of this project, then, is that connecting or reconnecting my students to nature needs to be one of my primary goals and considerations.

In order to help my students develop a stronger identity in relation to *Spatial Autonomy* and *Environmental Competency*, my future design work will carefully integrate experiences that create opportunities for students to experience the environment in positive ways. Such activities will include utilizing an outdoor classroom concept by connecting learning to aspects outside our classroom walls and spending time in our schoolyard habitat. Other activities will focus on creating opportunities to develop skills that promote a sense of joy and fulfillment through physical activity including hiking, climbing, and skiing in and around our school. Additionally, creating opportunities with students that allow them to creatively explore the natural world through drawing, painting, writing, composing, and photography may help foster feelings of reverence for our local context.

These types of intentional activities to enjoy nature place students outside and within the environment in presumably safe, low-risk situations to foster a positive emotional connection to the land. At the same time, emphasis will be placed on providing opportunities for students to encounter environmental dilemmas that pose a certain degree of risk and challenge. This, in turn, will not only force them to develop skills to navigate difficult situations but also encourage them to negotiate their own inner emotional tensions. For instance, during activities such as climbing and skiing, students will be encouraged to reflect on and work through situations that they find difficult or challenging (e.g., traversing a slippery slope). I hope that this will not only help students negotiate their feelings of *Mistrust in Nature* and *Environmental Shame*, it will also build up their confidence in the acquirement of *Spatial Autonomy* and *Environmental Competency*.

With that said, promoting a renewed love of and sense of comfort in nature is not enough. The design work of this project is foundationally grounded in student collaborative, cross-cultural inquiry. In considering climate change as a global issue, I hope that learning experiences will promote open discussions where students can respectfully share their feelings of *Environmental Disdain* and their fears that their actions will not make a difference. Lessons will engage students in analyzing case studies highlighting data from international, national, and regional mitigations to show that despite uncertainty, local actions can have positive impacts. Such lessons will be paired with discussions directed towards finding solutions to place-specific issues. These learning intentions are strongly supported by EID theory, promoting progression toward *Environmental Action*. It is critical that while engaged in inquiry investigations that students are simultaneously given opportunity to meta-cognitively engage in self-reflection about their emotional values, knowledge, and beliefs and how these affective elements influence a call to act. Initial opportunities to self-reflect and practice meta-cognition will support students in recognizing the context in which their actions and experiences are situated, specifically, by promoting a place-based educational orientation (Smith, 2013).

As well, *The GLOBE Program: Global Learning and Observations to Benefit the Environment* will be utilized to jointly engage Finnish and Alaskan students together around citizen science investigations in regards to global climate change (NASA, 2018). This worldwide program provides an interface for primary and secondary students to collect and analyze scientific data on their local environment. As a citizen science program, students engage as active researchers to collect, analyze, and interpret changes, which may be occurring in their local ecology. By promoting inquiry and transnational dialogue, students can compare and contrast local changes with those that are occurring in other parts of the globe. In this way, the project design includes opportunities for education in, for, and about the environment leading to *Environmental Action* and participatory methodologies that engages students in creating change.

Helping students understand that our world is a common community shared with others is an essential value in this educational project. Additionally, the project is designed to promote social interactions with same-age peers from a different part of the globe, which has been shown to have a significant impact on EID (Blatt, 2014). As Carie articulates in the introduction of this book, "it is also important to recognize that an individual's environmental identity develops in diverse ways and is highly influenced by family, sociocultural, and geographical contexts." The design of this project is to intentionally extend the boundaries of influences on my students' EID by increasing their sociocultural and geographical contexts through promoting transnational dialogue. This powerful aspect of engaging with peers of similar age is necessarily created in partial recognition that young learners at any age are continuously positioning their self-beliefs about knowledge and learning in relationship to interactions with their peers. By extending their community of learners, this project work will broaden their influences in potentially positive ways.

Influenced by the EID model, this project will specifically target international connections with peers residing in the global Arctic in negotiating *Environmental Competencies*. This will be done through providing opportunities for students to compare their environmental contexts and sharing feelings and emotions related to their environmental experiences. One of my observations while living in Finland for seven months in 2014 was that somehow our climates (Alaska and Finland) shaped people similarly in their environmental identity and relationship to nature. These similarities seem also to extend to cultural interactions and behaviors, exhibited in adults in both populations, such as strong personal determination demonstrated in difficult or adverse situations, a conscious limit on the necessity of communication and words, the restorative reconnection to our inner selves by escaping to nature, and a strong ancestral identity based on connection to the land and its resources. It is less clear whether younger generations are trans-generationally experiencing the same interactions and behaviors in relationship to the environment as their parents and elders. Through creating opportunities for my students to meta-cognitively self-reflect while engaging in learning experiences in nature, I hope that we can begin to understand and articulate enduring cross-cultural similarities that we share in relationship to our identities influenced by our Arctic global environmental context. These common understandings of natural, environmental, and cultural attributes may potentially be critical factors in navigating controversial and difficult global conversations between countries about climate change—conversations that our students will inevitably facilitate both now and in the future.

Bridging Cultures: Environmental Identity Development in Thailand and Alaska

By Robin Child, University of Alaska Fairbanks

Introduction

Using the EID framework proposed in this book, I investigated the factors that shaped two of my cousins' relationships with the natural world. Noah[1] is 10-years-old and is from Thailand; Eric is 11-years-old and is growing up in

Anchorage, Alaska. I have known these individuals for most of their lives and have had ample time to observe them in the familiarity of their households as well as other unique outdoor situations such as camping trips, family get-togethers, hikes, and road trips. Through observations and conversations with my cousins, I gained insight into the various factors that have influenced their environmental identity formation. Similarly, by comparing the geographical locations of their homes, their family socio-economic status, culture, and their parents' backgrounds, I note that both children exhibit a unique environmental identity. My observations also raised questions about how different cultures approach EE, and other influencing factors in EID.

Noah

Noah, my cousin and godson, was born and raised in Chiang Mai, Thailand. His father, my uncle Richard, passed away from cancer five years ago, and his mother, Mary, has raised him by herself in Doi Saket, a suburb north of Chiang Mai surrounded by rice paddies. As an only child of mixed race, Noah has grown up in a very unique situation in comparison to most Thai children, resulting in a distinct environmental identity. First off, his father, an American, came from a family that is notoriously obsessed with the outdoors. As a young child, Noah was taken camping and hiking, types of recreation that are not a part of Thai culture. Mary grew up in a rural part of southern Thailand, the daughter of very poor rubber tree farmers. She, too, was raised very closely with nature, waking in the middle of the night when temperatures were cool to go cut the trees to "bleed" rubber, walking miles to school everyday, knocking fruit from trees right outside the door for food, and bathing in the nearby river. She is the youngest of 10 and the first of her family to go to college. Her strong environmental identity grew not from recreation, then, but necessity and way of life, although Richard passed on his love for these outdoor activities to Mary who has continued to take Noah hiking and camping after Richard's death. Between her and Richard, the seed was planted early on for Noah to develop *Trust in Nature*, and he was provided with consistent encounters with the natural world as an infant, as recommended in the first stage of the EID model.

Mary, a single parent who struggles under financial strain, has done an incredible job in raising Noah. Despite the location of their home within a densely populated city, she prioritizes giving him a good education and finding ways to allow him to enjoy nature. At their home in Doi Saket, they have a large backyard with a goldfish pond and a big flower garden, which she and Noah now plant together. As a toddler, Noah was free to roam and explore, building *Spatial Autonomy* within his environment while his parents looked on, enabling independence. As Noah grew older, he could explore in wider berths. The family often escapes the heat of the city by going to the mountains outside of Chiang Mai to hike. One of Noah's favorite spots is Nam Tok Buatong, the "Sticky Waterfall," where one can walk up through the water as if with feet of Velcro like a gecko. Above the waterfall there is a small lake filled with lotus flowers. Noah told me about this special place:

> *You can take the water from the very top and drink it; they say it's for good luck. We go to waterfalls to find new places to be with nature because it feels better than the hotness in the city. It feels peaceful, not in a rush. It feels like home.*

Through exploration of nature, Noah has developed *Environmental Competency,* which, in turn, informs his sense of self in the natural world. His statement, *"It feels like home,"* shows his *Spatial Autonomy* and his biophilia, or love towards the natural world (Kalvaitis & Monhardt, 2015). It also conveys how a special place like the waterfall provides Noah with a sense of belonging, freedom, and life support.

Noah and Mary are Buddhists as are most people in Thailand. Temples and monasteries are plentiful, and visiting these sacred places is a regular part of Noah's life. The teachings of Buddhism, which promotes care and mindfulness for the environment, contributes to Noah's environmental identity as does the lifestyle that Mary and Noah lead as Buddhists. Noah explained:

> *We go exploring in caves that monks used to be in. It feels cool [temperature] and there are old statues. We go there to pray and walk around. Sometimes we go do the meditation in the mountains with the lady monks. It is very quiet but it [doesn't] feel lonely because there are still people walking around you but not talking to you so much. In the morning, we walk together with the monk—a walking meditation to the rice fields, strawberry fields, and to the local village—for around 2 ½ hours.*

White (2008) discussed how childhood activities give "form to the values, attitudes, and basic orientation toward the world that [a child] will carry with them throughout their lives" (p. 3). Growing up in a culture that teaches the values of finding peace, patience, and connection within nature contributes greatly to Noah's environmental identity and lifelong relationship with the natural world. He exudes maturity and patience way beyond his years, and shows a gentle love for other living things, qualities I attribute to his religion.

Besides the support that Mary gives at home towards the development of his environmental identity, the private school that Noah attends also seems to provide opportunities for learning about the environment. This year, Noah's class planted the garden next to the school's soccer field and waters it weekly. The vegetables are used in the school's cafeteria for lunch. This project shows progress in Noah's development into the fourth stage of the EID model, *Environmental Action*. His class also goes on field trips to the zoo and the Queen's Garden, which is a temple surrounded by many different species of orchid. Noah explained that they visit each year in the wintertime and explore different places in the garden, recording what information they find on paper, an activity that promotes inquiry and interest in the natural world.

Between experiences provided at home as well as at school, Noah has grown to love being outside, playing in and learning from his surroundings. On one of his visits to Alaska when he was 7, I took him out on the Root Glacier in the Wrangell Mountains to crawl around in ice caves. The first time we picked our way down into a blue tunnel, he cried out of fear. Through this experience, Noah grappled with the tension between *Trust in Nature* vs. *Mistrust in Nature*. Noticing his apprehension, I helped him gain confidence and make his way into the caves by having him place each foot exactly where mine had been, step by step, until we got to a place where we could comfortably stand and enjoy the blue ice in awe. When his cousin Eric arrived the next day, Noah begged to take him to the cave, and was the one helping him into the tunnel, eyes shining bright with pride and wonder. This demonstrates that Noah had developed a sense of *Trust in Nature*, which led him to share his feeling of *Spatial Autonomy* in a new environmental context.

Eric

Eric, age 11, is the son of my cousins, Drew and Suzy. Eric, as well as his brother, Timothy, age 9, were born and raised in Anchorage, AK. Drew and Suzy are like older siblings to me and are a big part of the reason that I moved to Alaska. I have known their kids since they were babies and have spent a lot of time watching their identities development. Both of Eric's parents were raised in the bush, north of Talkeetna, and spent their childhoods partaking in typical homesteader's activities: gathering subsistence food, putting up wood, hauling supplies from the railroad tracks (before the road was built all the way through to Fairbanks), and later skiing or hiking in and out to the road to get to school. Both come from families of mountaineers and outdoor enthusiasts, and both have grown into hardcore mountain runners and enjoy recreating like many Alaskans: hiking, skiing, fishing, and camping.

Eric was raised in a house that his parents built on the hillside above Anchorage, just below Flattop Mountain in the Chugach. Flattop is an often frequented after-dinner "stroll" –straight up from their house to the summit– over a thousand feet of gain in less than a mile. This mountain is an important facet of Eric's backyard and of his family's fitness regime. Many trails surround their house, leading off into the mountains from a backyard bordered by birch trees and alders. Because Drew and Suzy are so active and it seems that their physical and spiritual wellbeing hinge on time spent outside, it is certain that, as a baby, Eric spent a lot of time outdoors and was provided with the "consistent encounters ... to see, hear, smell, and touch nature" Carie argues allows him to establish a strong sense of *Trust in Nature*, the first stage of the EID model.

Eric and his brother spend a lot of time playing outside of their house. His favorite activities are bouncing on the trampoline, hanging out in the playhouse, and practicing on the slack line. He explained, "*I also go explore around our property on our trails whenever I want to go walk around.*" Eric's description of his environment affirms his *Spatial Autonomy* and sense of individuality in exploring his space, as well as the competence and confidence of his skills for navigating and playing in his backyard. By having "unmediated opportunities for adventure and self-initiated play, exploration and discovery" (White et al., 2008, p. 3), Eric is gaining knowledge that inspires his interest for the environment.

Engaging in outdoor activities together is important in Eric's family. Eric described his "*usual*" activities:

> Usually, we go skiing, but this year we haven't because the snow is so bad. We go on hikes up Flattop, and mountains near that. We go on [Uncle] Pete's boat down in Seward to go fishing for silvers. We find blueberries up on Flattop and in the mountains nearby. Sometimes we go to our cabin in Talkeetna.

Along with taking trips around the state, they go visit Suzy 's family in New Zealand, who live on a farm. The family also took a trip to Thailand last winter to visit Mary and Noah. Eric told me that his favorite part of visiting Thailand was riding elephants as well as going to the bug museum to hold scorpions and caterpillars. When asked if he was nervous, he replied, "*It was a little scary at first, but then it was cool. Because I grew up with lots of pets it helped.*"

Like Noah, Eric experienced a sense of *Mistrust in Nature* when he encountered a new environmental experience. However, Eric's fear was easily overcome as he drew upon his *Environmental Competency* built on his experiences and familiarity with pets and animals. Eric has had a family pets ever since he was born: Tooth and Molly, the dogs; Angel the guinea pig; Avia the cat; three mice; three gerbils; a gecko; hamsters; birds; and fish. It is Eric's responsibility to care for the animals with help from his brother, feeding and giving water to the rodents and cleaning their cages. Sobel (2008) discussed how "animals play a significant role in the evolution of children's care about the natural world and in their own emotional development" (p. 29). Exposure to animals is one way that children can "[develop] an emotional connectiveness-empathy-to the natural world" (White et al., 2008, p. 4), which is an essential component of EE. Drew and Suzy, in supporting contact with other living things and allowing Eric to build relationships with various types of creatures, have given him the ability to empathize, a characteristic that has already grown into a "willingness to care for other creatures" (Sobel, 2008, p. 30). This diligence for other living things shows Eric's progression into the fourth stage of the EID model, *Environmental Action*. His environmental identity begins to fit within a larger social context, and he is able to apply what he knows to act responsibly for the environment.

In addition to his home life, Eric's school seems to be a great source of development in his environmental identity. He attends a high quality public school close to his home, which has a lot of support for hands-on educational experiences. Eric's class goes on many field trips. He told me that his class studied salmon at the beginning of the school year, taking a trip to the hatchery and dissecting a salmon at school. In and out of class, Eric loves to keep a sketchbook, especially when he travels, using it to write down activities, people and places, and to draw different experiences. Eric explained, "I remember once I went to Mexico—I drew pictures of all the animals I saw ... iguana, crab, puffer fish." By keeping a record of his encounters, Eric is strengthening his *Environmental Competency* and his "interrelational agency" with other plants and animals (Ritchie, 2014). All and all, Eric's self-initiated curiosity for the natural world and the ease in which Eric navigates and takes care of his environment supports the conclusion that he is developing a strong sense of environmental identity.

Bridging Cultures

The week that Eric, Noah, and their families visited Unalakleet, Alaska, where I teach, winter finally showed its face, with temperatures dropping below −40°F with the wind. We spent most of our time outside anyways, skiing around "The Dragon's Back" (the snow drifts behind town), skijoring with the two dogs up and down the river, and picking up driftwood on the beach. We stood long hours outside at the Iditarod checkpoint, eagerly awaiting the arrival of dog teams, watching mushers feed and bed the dogs down. The boys stayed warm by climbing up and down the mounds of snow on the slough and sliding around on their bellies on the re-frozen overflow. The presence of a healthy environmental identity was obvious in my young cousins, simply from the enjoyment that each boy showed at being outside. Despite the weather, both boys have been conditioned to find enjoyment in the activity at hand. While Eric and Noah have had very different childhoods in their separate countries, households, and cultures, it is apparent that despite their different backgrounds and the various types of nurturing activities provided by their parents, each boy has come to see himself as a part of nature.

White et al. (2008) emphasized the importance of nurturing a child's environmental experiences. The parents of both boys model an enjoyment and love of nature, as well as support opportunities for their children to have both formal (e.g., school field trips, family trips, meditation retreats) and informal (e.g., exploring caves and waterfalls, bouncing on the trampoline, going berry picking, holding scorpions) experiences. These types of experiences, along with opportunities to travel internationally, build upon the stages of the EID model, continuing to strengthen their sense of *Trust in Nature*, *Spatial Autonomy*, and *Environmental Competencies*. These attributes, in turn, condition Eric and Noah's disposition to act responsibly for the environment.

While both boys exhibit characteristics that show *Environmental Action*, there are some differences in how comfortable each feels in solving environmental challenges. This is both related to the sociocultural and geographical context in which each is being raised as well as the scale of the environmental problems noted. Eric brought up how it was important to turn off the lights when he is not using them. Until last year, Eric's house was "off the grid" so the family had a generator and inverter to provide electricity when it was needed. Because of this firsthand experience, Eric is conscious of energy usage. Noah, on the other hand, shared this major environmental concern: "*This time of year, they are burning down the forests and rice paddies, so the air is very bad.*" In March, when the majority of the burning starts, the air quality in Thailand is so bad that those who can afford to leave the area until it subsides, including Noah and his mom. Others suffer through, wearing facemasks to fend off the smog. The environmental challenge that Noah faces is a very real problem, affecting thousands of people. When probed about how the problem could be fixed, Noah sighed and shrugged, seeming discouraged about finding a solution. I wonder about the presence of ecophobia, a fear of ecological problems in the natural world, in his 10-year old mind, and whether or not this feeling of powerlessness might lead to "apathy and reluctance to engage in environmentally responsible behavior" (Sobel, 1996, p. 8). I also wonder what he is being taught in school regarding this prominent environmental issue.

The differences in Eric and Noah's thoughts about environmental problems and solutions, as well as their sentiment towards taking action, raised questions for me about environmental identity and education. Does socioeconomic status affect an individual's feeling of powerlessness? How does the quality of EE differ between a private Thai school and a public American school in terms of the teaching of solutions vs. abstract problems? How does culture and religion play a part in an individual's EID and commitment to *Environmental Action*?

In sum, it is reassuring to know that reverence for nature has been cultivated in my cousin's lives from a very young age, and I feel lucky to have played a small role in promoting their EE. This analysis of my cousins' EID brings to the forefront of my own teaching practices an awareness of the role that culture plays in the rural communities of Northwest Alaska where I teach. In the Arctic North, children's identities are largely shaped by time spent outdoors engaging in subsistence practices. It helps me to think more critically about how I can best support the development of the environmental identities of my students within the unique cultural context of their daily lives.

Cultural Identity Development as it Relates to Environmental Identity Development

By Angela Lunda, University of Alaska Fairbanks

The lens through which I contemplate the complex concept of environmental/place identity development is that of an Indigenous Alaskan. For a Tlingít person, one's identity is intricately intertwined with *place*. When a Tlingít

person introduces herself, she shares her name, her clan, her ḵwáan (traditional geographic community from which her mother's people originate), and
the name of her house within that ḵwáan. She shares the same information for
her mother and grandmother's people as well as her father and grandfather's.
Names are clan at.óow (clan-owned property that can include intellectual
property such as names, stories, or songs, or material property such as blankets,
hats, or other regalia) that are recycled through generations; they carry stories, history, and are linked to specific places. By the end of the introduction,
the audience will be able to accurately situate the speaker in both spatial
and social contexts. Environmental/place and cultural identity are important
aspects of overall identity development that are nurtured in early childhood
through play and productive pursuits in nature as well as through stories that
support strong place attachment and deep understandings of place.

From an Indigenous worldview, people and place are one and the same,
for without a connection to the land, a person would be lost, rudderless
(Thornton, 2008). Place—the rocks, the trees, animals, the wind—are all
aspects of the environment we inhabit and are imbued with spirit. Respect
and stewardship for the land, water, animals, people, and the entire environment are values critical to the traditional subsistence lifestyle. When one's
survival depends on a healthy return of salmon or herring, then it is in the
best interest of the entire clan to ensure that the resource is conserved, that
no waste is tolerated, and that the environment supporting the resource is
healthy.

Tlingít values are taught and reinforced through subsistence practices and
stories. These cultural values were impressed upon me from my earliest years.
My family spent two to three months every summer at a remote fish camp on
the Taku River, a full day's travel by boat from our home in Juneau. My father
was a commercial salmon fisherman so he, along with my older brothers, spent
several days each week out on the fishing grounds leaving my mother, my
sisters, younger brother, and me at our two-room cabin. At our fish camp, we
caught salmon to smoke and can for use in the winter. We picked berries and
made jams to liven up the peanut butter and jam sandwiches that would grace
our lunchboxes most of the days during the school year back in Juneau. Without electricity or running water, and with wood as the only source of heat to
prepare food, fuel the fire in the smokehouse, or warm our cabin, we spent a
great deal of time working together to gather and cut firewood and collecting
fresh water from the nearby waterfall. Collaborative productive pursuit is one
of the most important strands in the development of Tlingít identity.

My mother taught us, by her example, the importance of respecting the resources that sustained us. She used every edible part of each fish, care-fully smoking the backbones and tails of the salmon, as well as the fillets; she removed the rich salmon eggs and canned them to add to soups during the winter months. She expressed gratitude for all of the resources that made themselves available to her large family. She implored the black bears who came to investigate the fish in the smokehouse on occasion, in Tlingít, to "allow us to take a few fish; we will take only what we need and leave plenty for you." We co-existed with the wildlife all of the years we spent the summer months at our Taku River fish camp. Derr (2002) describes the importance of these family-scale experiences of place as providing the historical and cultural context for experience ultimately allowing broader cultural values and place relations to take place.

Playing in and exploring the wild places near our cabin provided pivotal experiences in my EID. The mudflats provided the gymnastics mats where my siblings and I tumbled and slid, stopping to investigate a clump of grass hiding a hermit crab or other small creature. The moss-covered logs and lichen-strewn trees became castles with comfortable sitting areas decorated with intricate tapestries. The many days spent exploring the natural world contributed to my naturalist intelligence (Gardner, 2006), which led me to study biology and oceanography and contributed to my strong sense of stewardship for our planet. My interest in the natural world was born out of my many days spent exploring the tide pools, forests, meadows, and beaches of my childhood, but more importantly, it was the collectivist nature of our work that instilled in me the strong sense of attachment to place.

Our family caught all five species of salmon and each had its own special preservation technique. Dog (chum) salmon and humpy (pink) salmon were cut thin for drying while the richer coho (silver) and sockeye (red) salmon were cut into strips for smoking and then canning. King (Chinook) salmon, the least abundant, were reserved primarily for eating fresh as a succulent treat during fish camp. Everyone in our family was involved in the catching and processing of salmon. From the time we could hold a fishing pole, we fished in the slough near our cabin. When we caught a small Dolly Varden (Arctic char), our mother would show us how to cut it so we could make a miniature dry fish that looked just like one of the bigger dog or humpy dry fish. She would help us carefully hang this fish in the smokehouse and we would proudly check it everyday, awaiting the day when we could share it with our family. When we traveled by small boat up the river to pick salmon berries or nagoon berries, we were given small berry cans, just like the larger ones that our mother and father used. We were celebrated when we poured our full can of berries into the big pot that would soon be bubbling on the woodstove on the way to becoming delicious jam. In these ways, we became producers of food that contributed to our family wellbeing and to our identities as Tlingít people.

Through this traditional Tlingít lifestyle one can clearly see the development of cultural identity so important to *haa kusteeyí* (literally, our Tlingít way of life), referring to the intricate and intimate subsistence and spiritual relationship clans share with specific geographic locales. The traditional subsistence lifestyle also illustrates the parallel progression through the stages of EID as described by Carie in this book. When the bears approached the smokehouse, our mother's reaction was to speak calmly to them to reassure them, and us, that we could coexist peacefully. Sometimes she would remind the bears that she, too, had "cubs" to feed but that there was plenty of fish for both families, the bears and the people. In this way, she helped us develop *Trust in Nature*. Her actions further reinforced the concept that animals have parity with humans; she did not seek to exercise dominion over the bears, rather she talked with the bears just as she would talk with another person. When we were old enough to venture out away from the cabin on our own, we moved through *Spatial Autonomy*, developing our own sense of place, and into *Environmental Competency*, where we were allowed ample opportunities to experiment and solve problems through dramatic play. Our play was completely unstructured by adults; we sought out our own special places in the woods and tide flats around our Taku River cabin, creating imaginary worlds and constructing forts and other structures with tree limbs, fallen logs, and other "loose parts" available in the wild area we were fortunate to occupy. When we began contributing to the food stores by catching fish and picking berries, we moved into *Environmental Competency* and *Environmental Action*, exhibiting skills and confidence in our interaction with the environment and concomitantly developing stewardship values toward the valuable resources and the intricately interconnected world in which they are situated.

Because Tlingít had no written orthography until after the first contact with Russians, stories, songs, and art became the "books" that carried important cultural memories from one generation to the next. Some of these stories were important in environmental and cultural identity development. Indigenous people throughout the Northwest coast tell one such story[2] known as the "salmon boy" story, or "the moldy end" (Swanton, 1909, pp. 301–310).

Briefly, this story tells of a young man who, when offered a piece of dried salmon collar by his mother, scoffed at the piece of salmon and tossed it away because it had mold on it. The boy was taken under the blanket of the ocean and carried away by the Salmon People to live with them for a year. While he was with them, he learned about the ways of the Salmon People, including how they wished to be treated when they chose to give themselves to a person for sustenance. The boy became wise in the ways of the Salmon and returned to the beach where his mother was fishing. She caught the boy in the form of a salmon and knew it was the son she had been mourning for an entire year by the copper necklace he was wearing around his neck. After some ceremonies were performed, the boy returned to the form of a human and went on to become a powerful shaman. He taught his people the ways of the Salmon including the most important lesson of all: to be respectful of salmon from the moment they are pulled from the water, after giving themselves to the people, until they are consumed. This story is shared with young children, older children, and even adults and, because of the oral tradition of the Tlingít, the story can be tailored in style and tone to adapt to the developmental level of the audience. Different aspects of the story can be emphasized depending on the lesson the teller wishes to impart. For younger children, the story might emphasize the disrespect the boy shows the salmon collar in order to teach children the importance of not wasting food. For older children, the oral rendition might emphasize the specific lessons about the care and handling of salmon that the boy shared with the people upon his return to human form. The deepest interpretation of this story illustrates the intimate connection of people and place. The story acknowledges that the Salmon People have their own social structures that the boy discovers when he lives with them for a year. The story also emphasizes the complexity of this Salmon society as it took the boy an entire year to learn the important lessons that the Salmon wished to teach him. Finally, the salmon boy story demonstrates the spiritual link between people and animals through the transformation of the boy into a salmon. The line that separates people from salmon is seen as permeable; the boy simply had to slip under the blanket of the ocean to enter the world of the Salmon. Hearing the salmon boy story over many years, in many different iterations, helps to support the progression through the stages of environmental and cultural identity development.

Play and productive pursuits in nature, as well as stories such as "salmon boy," help to move children through the EID progression to *Environmental Action*, where children develop values and knowledge of place that can be applied toward environmental stewardship as Carie describes in this book. Parallel to the EID progression and in response to similar experiences, Tlingít children move through a progression toward cultural identity development. *Haa shagoon* is a core traditional Tlingít value that encapsulates this pinnacle of EID progression, *Environmental Action*. *Haa shagoon* speaks to the destiny (of a person or clan) stemming from the ancestors and extending out to future generations. A properly raised Tlingít child knows who she is, understands how she fits into the fabric of the physical and social landscape, and understands how she is responsible to steward the resources of *Lingít áani* (homeland) for future generations.

A Final Note

An individual's EID is never-ending. The mysterious uncertainty of our changing environment is what calls us to explore, to make connections, and to just be in nature. However, this call to the wild also presents a vast array of environmental challenges and personal tensions that we must overcome. Indeed, such tensions are inevitable in a quest to live life to the fullest and to be fully aware of the world around us. Writing a book on EID and considering the diverse ways in which a child's identity is formed has challenged me to evaluate my own strengths and weaknesses, my assumptions and biases, and to question the larger constructs of our contemporary society that promulgate an anthropocentric mindset even in the very young. There is much to be learned from a return to the basics, that is, the dispositions and skills necessary for survival.

Further, embedded in human's basic connection with nature is a spirituality that runs so deep that it awakens an essence of belonging that is undeniable. It is a way of seeing, a way of believing, and the harkening of an inner conviction that something is not right and that humanity must change if we and all with whom we share the planet are to flourish. Trust seems to be at the heart of this change, opening up questions of how humans can trust nature, whether nature can trust humans, and how we might trust each other. We have conquered, stolen, oppressed, desecrated, and trampled others, including the land that breathes under our feet. We are called to reconciliation, to humbled acknowledgement, and to an inevitable duty that something must change. We cannot keep forging forward without looking back nor can we truthfully be in the present without recognizing the plight of others and the world around us. These very things cut to the core of an identity crisis and to a crisis of human survival. Children are our hope and our future. This book aims to help us to critically consider both what and how we nurture our children to be in, with, and for their environments.

Notes

1. Names of the children and adults are pseudonyms.
2. Various Northwest coast clans own versions of this story. Out of respect for the ownership rights, this brief synopsis does not include any names of people or place-names that would tie it to a particular clan. Please note, then, that the omission of the names and place-names diminishes the power of the story and the does not do justice to the deep connection to place evident in the full versions of the story (see Swanton, 1909, pp. 301–310 or Thornton, 1985, pp. 73–80).

References

Blatt, E. (2014). Uncovering students' environmental identity: An exploration of activities in an environmental science course. *The Journal of Environmental Education, 45*(3), 194–216.

Clayton, S. (2003). Environmental identity: A conceptual and an operational definition. In S. Clayton & S. Opotow (Eds.), *Identity and the natural environment* (pp. 45–65). Cambridge, MA: MIT Press.

Derr, V. (2002). Children's sense of place in Northern New Mexico. *Journal of Environmental Psychology. 22*(1–2),125–137.

Fund for Teachers (2013). Retrieved from http://www.fundforteachers.org/

Gardner, H. E. (2006). *Multiple intelligences: New horizons*. New York, NY: Basic Books.

Kalvaitis, D., & Monhardt, R. (2015). Children voice biophilia: The phenomenology of being in love with nature. *Journal of Sustainability in Education*.

Kawagley, A. O. (2006). *A Yupiaq worldview: a pathway to ecology and spirit*. Long Grove, IL: Waveland Press.

National Aeronautics and Space Administration (NASA). (2018). *The Globe Program: Global Learning and Observation to Benefit the Environment*. Retrieved from https://www.globe.gov/.

Sadler, T., Klosterman, M., & Topcu, M. (2011). Learning science content and socio-scientific reasoning through classroom exploration of global climate change. In T. D. Sadler (Ed.), *Socio-scientific issues in the classroom: Teaching, learning and research.* (pp. 45–77). Netherlands: Springer Science.

Scannell, L., & Gifford, R. (2010). Defining place attachment: A tripartite organizing framework. *Journal of Environmental Psychology, 30*(1), 1–10.

Smith, G. (2013). Place-based education: Practices and impacts. In R. B Stevenson, M. Brody, J. Dillon, & A. E. J. Wals (Eds.), *International handbook of research on environmental education* (pp. 213–220). New York: Routledge.

Sobel, D. (1996). *Beyond ecophobia: Reclaiming the heart in nature education.* Great Barrington, MA: Orion.

Sobel, D. (2008). *Children and nature: Design principles for educators.* Portland, ME, Stenhouse Publishers.

Strife, S. (2012). Children's environmental concerns: Expressing ecophobia. *The Journal of Environmental Education, 43*(1), 37–54.

Swanton, J. (1909). *Tlingít myths and texts. Bureau of American Ethnology,* Bulletin 39. Washington, D. C.: Government Printing Office.

Thornton, T. (2008). *Being and place among the Tlingít.* Seattle, WA: University of Washington Press.

White, R., & Stoecklin, V. L. (2008). *Nurturing children's biophilia: Developmentally appropriate environmental education for young children.* Kansas City, MO: White Hutchinson Leisure and Learning Group.

Zeyer, A., & Kelsey, E. (2013). Environmental education in a cultural context. In R. B. Stevenson, M. Brody, J. Dillon, & A. E. J. Wals (Eds.), *International handbook of research on environmental education* (pp. 206–212). New York, NY: Routledge.

CONTRIBUTOR BIOGRAPHIES

Robin Child lives in Unalakleet, AK and is the district-wide itinerant arts integration facilitator for the Bering Strait School District. She works with students from preK through 12th grade to bring the arts into the everyday classroom, helping students forge connections between core content areas- reading, writing, social studies, math, and science- through hands-on, creative experiences. She strives to draw upon the rich cultural knowledge, subsistence tradition, and unique landscape of the Bering Strait region as a foundation for teaching in and through the arts. Child may be contacted at child.robin@gmail.com.

Angela Lunda, *Ḵoogak'aax̱*, currently teaches in the Secondary Master of Arts of Teaching program at the University of Alaska Southeast and she is a student in the Indigenous Studies PhD program at the University of Alaska Fairbanks. She is a life-long Indigenous Alaskan of the Tlingít tribe, *Ch'aak* (Eagle) moiety, *Kaagwaantaan* (Wolf) clan, and the Sitka *Déix̱ X'awool'ja Hít* (Two-Door House) with more than three decades as a teacher and administrator. Lunda may be contacted at angielunda@gmail.com.

Karen Martin is a 4th Grade Teacher and Instructional Coach for Denali Borough School District in Alaska, with 13 years experience. Karen holds a Master of Arts in Teaching and a Master of Science in Genetics. She was a recipient of a *Distinguished Fulbright Awards in Teaching* grant (2014) and a *Fund For Teacher Fellow* (2017), which allowed her to study education in Finland. She is currently pursuing an Interdisciplinary Ph.D. at the University of Alaska Fairbanks with a focus on professionalizing teaching through action research. Martin may be contacted at karenjoannmartin@gmail.com.

GENERAL EDITORS: CONSTANCE RUSSELL & JUSTIN DILLON

The [Re]thinking Environmental Education book series is a response to the international recognition that environmental issues have taken center stage in political and social discourse. Resolution and/or re-evaluation of the many contemporary environmental issues will require a thoughtful, informed, and well-educated citizenry. Quality environmental education does not come easily; it must be grounded in mindful practice and research excellence. This series reflects the highest quality of contemporary scholarship and, as such, is positioned at the leading edge not only of the field of environmental education, but of education generally. There are many approaches to environmental education research and delivery, each grounded in particular contexts and epistemological, ontological and axiological positions, and this series reflects that diversity.

For additional information about this series or for the submission of manuscripts, please contact:

Constance Russell & Justin Dillon
c/o Peter Lang Publishing, Inc.
29 Broadway, 18th floor
New York, New York 10006

To order other books in this series, please contact our Customer Service Department:

(800) 770-LANG (within the U.S.)
(212) 647-7706 (outside the U.S.)
(212) 647-7707 FAX

Or browse by series:

WWW.PETERLANG.COM